CHEMICAL ENGINEERING: I

DIMENSIONS OF SCIENCE

Practical Ecology D. Slingsby and C. Cook
Ionic Organic Mechanisms C. Went
Chemical Science in Conservation D. Burgess
Physics and Astronomy P. McGillivary
Electrolyte Solutions M. Wright
Energy and Cells C. Gayford
Human Reproduction and I.V.F. H. Leese
Chemical Engineering: Introductory Aspects R. Field
Science and Criminal Detection J. Broad
Real Applications of Electronic Sensors G. Long
Genes and Chromosomes J. Lloyd

DIMENSIONS OF SCIENCE
Series Editor: Professor Jeff Thompson

CHEMICAL ENGINEERING
Introductory Aspects

Robert W. Field

University of Bath

© Robert W. Field 1988

All rights reserved. No reproduction, copy or transmission of this publication may be made without written permission.

No paragraph of this publication may be reproduced, copied or transmitted save with written permission or in accordance with the provisions of the Copyright, Designs and Patents Act 1988, or under the terms of any licence permitting limited copying issued by the Copyright Licensing Agency, 90 Tottenham Court Road, London W1P 9HE.

Any person who does any unauthorised act in relation to this publication may be liable to criminal prosecution and civil claims for damages.

First published 1988 by
THE MACMILLAN PRESS LTD
Houndmills, Basingstoke, Hampshire RG21 2XS
and London
Companies and representatives
throughout the world

ISBN 0-333-45249-6

A catalogue record for this book is available from the British Library.

Printed in Hong Kong

Reprinted 1992

Series Standing Order

If you would like to receive future titles in this series as they are published, you can make use of our standing order facility. To place a standing order please contact your bookseller or, in case of difficulty, write to us at the address below with your name and address and the name of the series. Please state with which title you wish to begin your standing order.
(If you live outside the United Kingdom we may not have the rights for your area, in which case we will forward your order to the publisher concerned.)

Customer Services Department, Macmillan Distribution Ltd, Houndmills, Basingstoke, Hampshire, RG21 2XS, England.

*To Catherine
and our children
Marianne, John and Katy*

Contents

Series Editor's Preface   ix

Preface   x

Acknowledgements   xi

1 **Chemical Engineering in Context**
Chemical engineering — the subject. The process dimension. The challenge and rewards of chemical engineering. Case study 1: Sulphuric acid production. Case study 2: Production of ammonia. Case study 3: Process engineering in the food industry. Case study 4: Process engineering off-shore. Case study summary. Exercises   1

2 **Process Design**
Material balances: general. Material balances: unsteady state. Material balances: with reaction. Material balances: recycles and purge streams. Material balances: a summary. Energy balances: general. Energy balance equation. Process economics. Exercises   25

3 **Fluid Flow**
Newtonian and Non-Newtonian fluids. Nature of fluid flow. Pressure drop in a pipe. Exercises   52

4 **Heat Transfer**
Thermal conductivity. Heat loss across windows: an oversimplification. Radial heat flow. Heat transfer coefficients, U-values and fouling factors. Temperature driving force. Sizing and types of heat exchanger. Radiation. Energy recovery. Exercises   62

5 Separation Processes
Range and choice. Case study 1 revisited. Case study 2 revisited. Case study 3 revisited. Case study 4 revisited. Membranes and membrane processes. Process selection. Exercises   **91**

6 Distillation and Absorption
Factors affecting distillation column design. Minimum reflux ratio. Minimum number of stages. Balancing reflux ratio against number of stages. Choosing and sizing trayed columns. Packed columns compared with trayed columns. Absorption. Exercises   **113**

7 Chemical Reactors
Batch or continuous operation. Continuous reactors. Plug flow reactor. Continuous stirred tank reactor. Case studies revisited. Case study 1: sulphuric acid production. Case study 2: ammonia production. Case study 3: related to the food industry. Case study 4: process engineering off-shore. Exercises   **136**

8 Process Control and Safety
Feedback control. Process safety: an introduction. An approach to safer design. Conclusion. Exercises   **158**

Glossary of Terms   *174*

List of Symbols   *177*

Index   *179*

Series Editor's Preface

This book is one in a Series designed to illustrate and explore a range of ways in which scientific knowledge is generated, and techniques are developed and applied. The volumes in this Series will certainly satisfy the needs of students at 'A' level and in first-year higher-education courses, although there is no intention to bridge any apparent gap in the transfer from secondary to tertiary stages. Indeed, the notion that a scientific education is both continuous and continuing is implicit in the approach which the authors have taken.

Working from a base of 'common core' 'A'-level knowledge and principles, each book demonstrates how that knowledge and those principles can be extended in academic terms, and also how they are applied in a variety of contexts which give relevance to the study of the subject. The subject matter is developed both in depth (in intellectual terms) and in breadth (in relevance). A significant feature is the way in which each text makes explicit some aspect of the fundamental processes of science, or shows science, and scientists, 'in action'. In some cases this is made clear by highlighting the methods used by scientists in, for example, employing a systematic approach to the collection of information, or the setting up of an experiment. In other cases the treatment traces a series of related steps in the scientific process, such as investigation, hypothesising, evaluating and problem-solving. The fact that there are many dimensions to the creation of knowledge and to its application by scientists and technologists is the title and consistent theme of all the books in the Series.

The authors are all authorities in the fields in which they have written, and share a common interest in the enjoyment of their work in science. We feel sure that something of that satisfaction will be imparted to their readers in the continuing study of the subject.

Preface

Chemical Engineering: Introductory Aspects aims to reveal some of the links between A-level science and the distinct subject of chemical engineering. A further objective is to reveal some of the work done by chemical engineers, to illustrate a few of the benefits to society that have resulted, and hence to show some of the links between industry and the everyday products around us. It is anticipated that the book will be of interest and use to both students in school as well as those starting a university course. Terms printed in bold in the text are defined in the glossary.

I am grateful to Sally Barnet for typing from a none-too-legible manuscript. The help of Barry Crittenden and David Rowe, who read the manuscript and contributed suggestions, is also acknowledged with thanks. Any remaining errors and omissions are, of course, solely my own responsibility. Finally my thanks go to my wife and children for their patience and understanding during those periods when I was especially busy working on the text.

RWF

Acknowledgements

The author and publishers wish to thank the following who have kindly supplied photographs or diagrams.

APV UK Ltd   figures 4.9, 4.10 and 5.11
John Brown & Co Ltd   figure 1.11
Esso UK plc   figures 6.4 and 6.5
Glitsch (UK) Ltd   figures 6.12 and 6.13
M. W. Kellogg Ltd   figure 7.11
ICI Chemicals and Polymers Ltd   figures 1.6, 1.10 4.14, 7.10, 7.13 and 8.8
ISC, Avonmouth   figure 1.3

Every effort has been made to trace all the copyright holders but if any have been inadvertently overlooked the publishers will be pleased to make the necessary arrangement at the first opportunity.

1 Chemical Engineering in Context

CHEMICAL ENGINEERING — THE SUBJECT

Society can associate civil engineers with huge new building complexes and bridges, electronic and electrical engineers with telecommunications and power generation, and mechanical engineers with advanced machinery and automobiles. However, chemical engineers have no obvious monuments which create an immediate awareness of the discipline in the public mind. Nevertheless, the range of products in daily use which are efficiently produced as a result of the application of chemical engineering expertise is enormous. The list given in table 1.1 is not exhaustive, and any reader who grasps the key element, which involves the conversion of raw materials into a useful product, will be able to extend it. Although the products are unglamorous, the creation and operation of cost-effective processes to produce them is often challenging and exciting.

The term 'chemical engineer' implies that the person is primarily an engineer whose first professional concern is with manufacturing processes — making something, or making some process work. The adjective 'chemical' implies a particular interest in processes which involve chemical changes. While the main term is correct, the adjective is too restrictive and the literal definition will not suffice. Taken at face value, it would exclude many areas in which chemical engineers have made their mark. For example, textiles, nuclear fuels and the food industry. Thus the Institution of Chemical Engineers defines **chemical engineering** as "that branch of engineering which is concerned with processes in which materials undergo a required change in composition, energy content or physical state; with the means of processing; with the resulting products, and with their application to useful ends." It is perhaps too presumptuous to insist that the term 'process engineer' should replace the term 'chemical engineer', and so the two will be used synonymously.

Table 1.1 A selection of everyday products whose manufacture involves the application of chemical engineering

Product grouping or production process	Some of the more familiar examples
Household products in daily use	Detergents, polishes, disinfectants
Health care products	Pharmaceuticals, toiletries, antiseptics, anaesthetics
Automotive fuels/Petroleum refining	Petrol, diesel, lubricants
Other chemicals in daily use	Latex paints, rubber, anti-freeze, refrigerants, insulation materials
Horticultural products	Fertilisers, fungicides, insecticides
Metals	Steel manufacture, zinc production
Polymerisation, extrusion and moulding of thermoplastics	Washing-up bowls, baths, insulation for cables, road signs, children's toys
Polymerisation, production and spinning of synthetic fibres	Clothes, curtains, sheets, blankets
Electronics	Raw materials, silicon, gallium arsenide, etchants, dopants
Fats and oils	Salad and cooking oils, margarine, soap
Fermentation	Beer; certain antibiotics such as penicillin, yoghurts
Dairy products	Milk, butter, cheese, baby food
Gas treatment and transmission	Gas for heating and cooking

It should also be noted that large-scale processes involving biological systems (such as waste water treatment and production of protein) fit the definition as well as traditional chemical processes such as the production of fertilisers and pharmaceuticals.

The work of chemical engineers will be examined by way of four case studies in the second part of this chapter, but to complete the definition, explicit mention of the concern that process operations be both safe and economic must be made.

A jocular, helpful, but very incomplete description is that "a chemical engineer is a chemist who is aware of money." Although this neglects many, if not most, aspects of a chemical engineer's training, it does illustrate one important facet of any engineer's work. When working on a large scale, the cost of equipment and raw materials are more important than the cost of manpower. While the research chemist might use aqueous potassium hydroxide to neutralise acids, because it is pure and readily available, the chemical engineer will specify a cheaper alternative, provided that it serves the same purpose. Two obvious substitutes are aqueous sodium hydroxide, which is available at less than a tenth of the cost, or calcium hydroxide, which is even cheaper, but harder to handle. In choosing between these cheaper alternatives, an engineer has to balance the cost of handling a slurry (calcium hydroxide is sparingly soluble) against the higher price of sodium hydroxide.

The Process Dimension

The distinctive feature of chemical engineering (which some believe should be called process engineering) is the 'process dimension': the ability to break down into its component parts a manufacturing process in which matter is transformed or chemically changed, provide a specification for each subdivision and recombine the whole into a scheme or **flowsheet** which represents an economical, workable and maintainable plant.

Chapter 2 concentrates on Process Design, but it will be readily appreciated that all processes can be divided, as shown in figure 1.1, into at least three stages:

> Preparation — including purification of raw material
> Reaction or transformation
> Treatment of products, including separation and storage

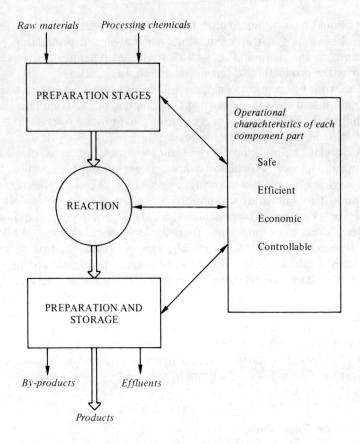

Figure 1.1 *Chemical engineering: a generalised process*

The component parts include processes such as the separation of liquid mixtures by **distillation** or **liquid–liquid extraction**, and of solids from liquids by filtering, **sedimentation** or centrifuging; the heating, cooling and evaporation of liquids and the condensation of vapours; the drying of solids; and the chemical conversion of a continuous stream of solid or fluid material.

Much of the success of chemical engineering over the last 50 years can be attributed to this approach. By breaking down a process into a series of steps, or '**unit operations**', the chemical engineer often moves away from a product-specific operation to one which is general. The separation of a liquid from a solid might involve **filtration**, which is a unit operation. The chemical composition of the material that is being filtered will affect the materials of construction, but little else. The important factors are available pressure, size of filter and hydraulic resistance, which will need to be determined. None of these is specific to one product and so the mathematical analysis which, for example, can lead to a prediction of maximum filtration rate, is general. The same is true of the other areas, and so knowledge gained while developing one process has been used to improve completely different processes.

The process dimension is now seen as being crucial to the success of most manufacturing processes and it is perhaps no accident that the chemical and allied industries, which exemplify this approach, have been conspicuous by their success. Despite the poor performance of most of the UK manufacturing industry, the UK chemical and allied industries have for many years had a very healthy positive trade balance of over £1.5 billion.

The Challenge and Rewards of Chemical Engineering

Before concentrating on typical chemical engineering jobs, it is appropriate to note that few degree courses give a better general technical education than chemical engineering, and that a quarter of UK graduates now move immediately into non-process engineering posts. In the USA, the present position is that 60–70 per cent of graduating chemical engineers are finding jobs in the aircraft, automotive, food and pharmaceutical, electrical and electronics industries. In the 1970s, the job market of the USA was similar to that of the UK, and the chemical and petroleum industries used to hire 70–80 per cent of the graduates. While such a turn around is unlikely in the UK, the trend of increasing opportunities in non-traditional areas which started in the 1980s will continue.

Although many have seen chemical engineering solely as the discipline which enables material goods (for example, plastics, fibres, detergents, paints) and commodity chemicals (for example, fertilisers, sulphuric acid) to be produced efficiently on the large scale, the discipline is also relevant for those wishing to emphasise environmental protection, energy conservation, improved food production and better health care. A dichotomy is not proposed; the quality of life depends upon a balance, and like present-day society, the profession of chemical engineering understands this.

The rewards are more than salary and the opportunity to travel. Besides helping to produce products that the community needs, there is the satisfaction of originating and developing new ideas, and the excitement of turning some of them into reality.

CASE STUDIES

This book's main aim is to demonstrate how knowledge gained from a study of science and mathematics at 'A' level standard can be extended and applied to process engineering. In furtherance of this aim, four case studies will be used to link various aspects of the subject.

In choosing these case studies, a balance has been struck between the wish to reveal the diverse range of activities in which process engineers engage and the need to link the examples to familiar material. The four case studies are contextually introduced in the following pages. They are a useful, relevant and, hopefully, illuminating source of material both for the illustration of the selected technical subjects that are introduced later and for certain non-technical facets of the discipline.

Case Study 1: Sulphuric Acid Production

Before producing a product for sale, the manufacturer needs to be certain that there is a market and that the chance of a return on the investment is reasonable. Some uses, with notes which indicate the changing nature of the sulphuric acid market, are listed in table 1.2. Although the UK market for sulphuric acid is a mature one, with excess manufacturing capacity, it will, for illustrative purposes, be assumed that a company is about to invest in a new plant which is to employ an improved catalyst developed in-house.

Consider the following equations:

$$S + O_2 \longrightarrow SO_2 \quad (1.1)$$
$$SO_2 + \tfrac{1}{2}O_2 \rightleftharpoons SO_3 \quad (1.2)$$
$$SO_3 + H_2O \longrightarrow H_2SO_4 \quad (1.3)$$

They reveal that the raw materials are sulphur, oxygen and water, and that the second reaction is an equilibrium one. However, little else is revealed, and some important questions are:

(1) What are the sources and state of the raw materials?
(2) Are the reactions endothermic or exothermic and to what extent?
(3) Reaction (1.2) requires a catalyst. Over what range of temperatures is the catalyst active? Is the reaction pressure fixed?
(4) Can the temperatures and pressures for each of the other steps be specified?

Table 1.2 Uses and sources of UK sulphuric acid in the 1980s

Market	Approx. annual tonnage (tonnes)	Comments
Uses		
Phosphoric acid and phosphate fertilisers	800 000	A declining usage. Developing countries which have reserves of phosphate rock are increasing their fertiliser production and selling the finished product in preference to the raw material
Paints and pigments	400 000	A declining usage. Titanium dioxide is now increasingly made by the chloride and not the sulphate route
Cellulose films and fibres	210 000	Usage directly related to production of cellulose; acid is consumed during the extrusion process, which results in the regeneration of the cellulose and the formation of Na_2SO_4. Market now stable; it declined sharply when polyolefin films penetrated the packaging market and polyester and other synthetic fibres cut into the cellulose fibre market
Detergents and soaps	420 000	Relatively stable
Chemicals and plastics	175 000	Small but steady increase
Other uses	900 000	Main area of increase. However, production of HF for manufacture of refrigerants and propellants (for aerosols) may decline because of the concern that chlorofluorocarbons are damaging the ozone layer of the Earth's atmosphere
Sources	*Percentage*	
Domestic sulphur recovered from crude oil	11%	The anhydrite and pyrite (FeS_2) based plants closed during the 1960s and 1970s
Imported sulphur	83%	Two-thirds of the sulphur arrives as liquid sulphur
Recovered SO_2	6%	The smelting of sulphur-containing ores (such as ZnS) produces SO_2 which is recovered and converted to H_2SO_4, thus simultaneously giving a valuable by-product and reducing atmospheric pollution

(5) Can the heating/cooling requirements be matched? If not, what are the economic solutions?
(6) What are the limits on gaseous emissions?
(7) What are the likely hazards?

With regard to question (1), sulphur may be delivered molten, in which case it has to be kept above its melting point of 113°C, or as a solid. The oxygen used is not pure oxygen but simply air since the presence of nitrogen is perfectly acceptable. A simplified process flow diagram for the process is shown in figure 1.2.

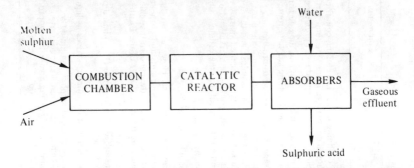

Figure 1.2 *Basic process flow diagram for sulphuric acid production: heat removal essential at every stage*

The research and development team of chemists and chemical engineers which developed the catalyst would be able to answer the third question. A possible answer would be: the desirable temperature range is 410–590°C with some activity at a temperature as low as 360°C, and severe damage above 620°C; the pressure will be just above atmospheric pressure and more attention is paid to this fact later. The answer to question (6) would be influenced by company policy and determined in conjunction with the regulatory authorities.

The design engineers within the company (mainly chemical and mechanical) will develop answers to all of the above questions, but because a full design will exceed 10 man-years, a firm or consortium of engineering contractors will often be used to complete the design. This will include not only the main items of equipment, but also the pipework, valves and pumps. In fact, they will be responsible for everything from the foundations to the last sample valve. Such work is at the heart of chemical engineering, and London is an important centre of the international contracting industry.

Close liaison between the company team and the contractors is vital. The reactions given above are all exothermic and economic use must be made of the excess energy. There is never just one answer, and what is economic for one company may not be so for another. Detailed consideration will be given to hazards such as fire, explosion or accidental release of fluids. The control system, which is often computer-based, will have to cope efficiently and effectively with abnormal conditions, start-up and shut-down, and the maintenance of the plant at an optimal level of performance. A sulphuric acid plant will run, unless problems arise, for 24 hours per day, 7 days per week, 50 weeks a year. The modern philosophy is to see control as an integral part of design and not an add-on. Thus chemical engineers have to be conversant with the subject of Process Control and for those wishing to specialise as control engineers, chemical engineering is a good first degree.

With the plant designed, **procurement**, construction, **commissioning** and start-up follow. These tasks are often undertaken by a contractor, generally the one who designed the plant. If a go-ahead was given early, some of the major items, such as acid-gas absorbers, will have been ordered before the detailed design was complete, because a 6 months wait for equipment is not uncommon. After erection, the inside of every vessel and pipe is cleaned and each item of equipment tested. Particular attention is paid to the alarms and safety trips. The commissioning team of engineers, with process engineers to the forefront, will work very long hours to bring the plant on-stream, as quickly as possible. The start-up of a sulphuric acid plant is reasonably straight-forward. The reactors have to be brought up to temperature before introducing sulphur dioxide, and they are heated slowly with hot dry air typically as follows: (1) from ambient temperature to 150°C at 10°C per hour, and (2) from 150°C to operating temperature at 25–30°C per hour. Thus as with other plants, the operation takes a matter of days rather than hours. A typical plant is shown in figure 1.3.

Once the new plant is running to specification, the commissioning team will hand over to the production staff and the contractor will have finished the task. The production staff will consist of production managers and process operators on the one hand, and plant engineers, technical managers and maintenance staff on the other. The former are responsible for making the product safely, economically and to specification. The latter will evaluate proposed longer-term improvements and aid with the '**trouble shooting**'. With a modern sulphuric acid plant, the process is sufficiently simple that on the production

Figure 1.3 *Sulphuric acid plant with reactor in foreground*

side, less than 15 operating staff will be able to provide 24 hour cover throughout the year.

The acid produced must be sold and the engineer may well move across to marketing and technical sales, which involve not only the selling of the product but also the provision of a technical service. Some engineers move into purchasing, finance, planning and related commercial departments.

Chemical engineering gives a broadly based technological training, and a chemical engineer can undertake a wide variety of tasks. Although subsequent chapters will concentrate upon technical subjects, such as heat transfer and separation processes, it should be remembered that many of the 75 per cent who are recruited into process engineering posts, move into management, where they have a certain advantage because chemical engineering is the key discipline which co-ordinates the activities of the process industries.

Case Study 2: Production of Ammonia

Ammonia is one of the most important chemicals being manufactured in large tonnage today. Unlike sulphuric acid, its production is not linked to the Industrial Revolution, but to the population explosion which followed. At the beginning of the twentieth century, there was a need to increase food production, and a necessary solution in the densely populated industrial world was to increase the yield per acre by use of nitrogen-containing fertilisers. Although nitrogen is generally unreactive, Fritz Haber showed that **thermodynamically** the reaction of nitrogen with hydrogen was feasible. In 1911 he discovered that an iron catalyst would facilitate production, and process development started. During the First World War, the manufacture of ammonia in Germany accelerated rapidly because it was a convenient home-based substance which could be converted to nitric acid, which was crucial to the production of explosives.

After the First World War, ammonia continued to be made by the Haber process, which was adopted in many countries, but the main outlet was in fertiliser production. World-wide production reached about 10 million tonnes per year in 1950, and is now nearly 100 million tonnes per year, of which 80 per cent is used to make nitrogenous fertilisers. The contribution that this makes to the nitrogen cycle (shown in figure 1.4) is very significant.

However, this increasing usage has not been problem-free. The levels of nitrate in drinking water are rising wherever intensive agriculture is practised, and although the actual health effects are in dispute, the World Health Organization and the EEC are laying down a maximum nitrate level of 50 mg of NO_3 per litre. A number of supplies currently exceed this level and the water industry is facing an unusual separation problem: the *selective* removal of nitrate ions from water.

Now the synthesis of many chemicals, particularly ammonia, is governed by chemical equilibrium considerations, and the effects of temperature, pressure and concentration upon the reaction are of paramount importance. In the past, many have sought to summarise the effect of temperature and pressure by application of Le Chatelier's deceptively simple (or, as some critics would say, simply deceptive) principle. A statement of the principle is given in the Glossary. A sound set of rules, based on Van't Hoff's principal laws of chemical

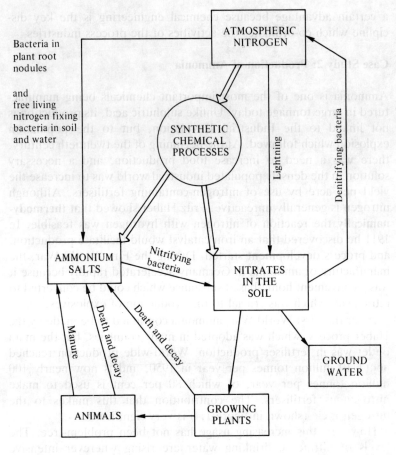

Figure 1.4 *The Nitrogen Cycle: man supplements soil-based nitrogen fixation by 75 per cent*

equilibrium are preferred. They are unambiguous, theoretically sound, and can be made quantitative:

(1) At constant temperature, increase in pressure favours the system possessing the smaller volume.
(2) Rise in temperature favours the system formed with absorption of heat.

With ammonia production there is both a reduction in the number of molecules and a significant evolution of heat.

$$N_2(g) + 3H_2(g) \rightleftharpoons 2NH_3(g) + \text{heat} \qquad (1.4)$$

Thus if the pressure is increased, the production of ammonia will be favoured. If the temperature of the system is reduced, the system formed with the liberation of heat will be favoured, and in this case, the equilibrium will shift to the right. So, in theory, the conversion will be greater if conditions of high pressure and low temperature are used (see table 1.3). However, in industry, rate of reaction is important, and a relatively low percentage conversion is accepted.

Table 1.3 Volume percentage of ammonia in equilibrium mixture for a feed with a mole ratio $H_2:N_2$ of 3:1

Pressure (bar)	Temperature (°C)			
	200	300	400	500
10	51	15	3.9	1.2
100	82	52	25	11
200	89	67	39	18
300	90	71	47	24
400	94	80	55	32
600	95	84	65	42

The unused reactants are recycled as shown in figure 1.5. Although this circumvents the need to have a high percentage of ammonia in the equilibrium mixture, the use of a recycle involves additional compressor costs and allows inert material to accumulate. Since the accumulation of inerts such as argon and methane has to be checked, it is essential to split the recycle stream into a main stream that is returned to the catalytic converter, and a small purge stream. The latter represents a loss. Recycles and purge streams are discussed further in chapter 2.

Equipment becomes increasingly expensive at high pressure, particularly if the temperature is also high, and so these and other considerations, such as the relative magnitude of the purge, result in a compromise. A typical pressure for plant built 10 years ago was 200 bar, but with the advent of more efficient catalysts (still iron-based) the trend is towards 75 bar. The temperature of the gases leaving the reactor is typically 400°C.

The development of the conversion process is a task shared by chemists and chemical engineers, but it is the chemical engineers with the help of other engineers, particularly mechanical engineers, who

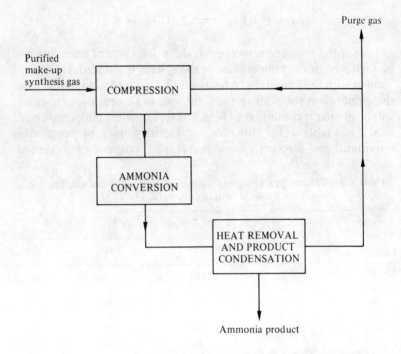

Figure 1.5 *Schematic flow diagram of ammonia synthesis: note importance of recycle*

design the complex plant that (as figure 1.6 shows) is clearly *not* a scaled-up piece of laboratory apparatus which would simply have a supply of pure hydrogen and nitrogen and no energy recovery systems.

In equation (1.4), nitrogen and hydrogen are the reactants, but from where do they originate? The answer is generally 'from natural gas, water and air'. The simplified flow scheme of a typical 1200 tonne/day ammonia plant based on natural gas is shown in figure 1.7.

After desulphurisation, a process which is necessary for the protection of the catalysts, the natural gas which is mainly methane is reacted with steam over a nickel catalyst. The reaction is overall endothermic, and so, in accordance with the laws of chemical equilibrium, as high a temperature as possible is required. The reactor, known as the primary reformer, is a collection of vertical metal tubes suspended in a furnace, and the exit gases around 800°C are unreacted methane 9 per cent, steam, oxides of carbon and hydrogen. The principal reactions occurring simultaneously are:

Figure 1.6 *Modern ammonia plant under construction*

$$CH_4 + H_2O \rightleftharpoons CO + 3H_2 - \text{heat} \quad (1.5)$$
$$CO + H_2O \rightleftharpoons CO_2 + H_2 + \text{heat} \quad (1.6)$$

In the next stage, the secondary reformer, air, is introduced. The principal aim is to remove the oxygen by combustion with some of the gas from the primary reformer. Another feature of this reaction is that the highly exothermic combustion raises the temperature substantially to about 1000°C, which has a desirable influence upon the equilibrium represented by the reaction (1.5). Ideally, the final gas composition would be a 1:3 ratio of nitrogen to hydrogen with carbon dioxide as an easily removable third component. However, some methane and carbon monoxide remain unconverted, and argon, which came in with the air, is also present. The processing steps taken to deal with these unwanted chemicals are discussed in chapter 7. However, before leaving this subject, it is worth examining the conditions in the methanator.

After removal of the vast bulk of carbon dioxide in the absorber, the synthesis gas stream of hydrogen and nitrogen still contains water, and a small amount of carbon dioxide and carbon monoxide,

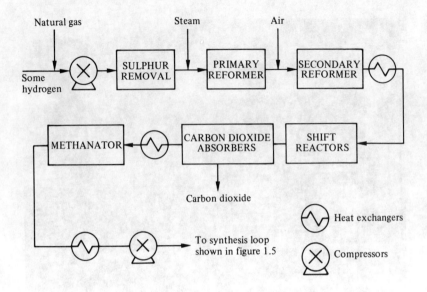

Figure 1.7 *Basic process flow diagram of an ammonia plant with natural gas feedstock*

all of which poison the iron catalyst in the ammonia reactor. The oxides of carbon are converted back into methane:

$$CO + 3H_2 \rightleftharpoons CH_4 + H_2O \qquad (1.7)$$
$$CO_2 + 4H_2 \rightleftharpoons CH_4 + 2H_2O \qquad (1.8)$$

Equations (1.7) and (1.8) should be compared with the reverse of equations (1.5) and (1.6). The methanation is favoured by the high hydrogen to oxides of carbon ratio now prevailing, and the much lower temperature (350°C) compared with that in the primary reformer. The resulting CO/CO_2 concentration is 1–2 parts per million.

At this stage, the reader should compare the **unit operations** of the ammonia and sulphuric acid processes. Although some processes have special requirements, it should already be clear that the chemical engineer can apply his experience of designing heat transfer equipment for ammonia production to similar sorts of problems in nylon, drug or refrigerant production. The links with the food industry is the next subject.

Case Study 3: Process Engineering in the Food Industry

The pumping of fluids, the pneumatic conveying of solids and many other preparatory steps are common to both the chemical and the food industries. Processing operations such as mixing, heating/cooling and reaction are in both areas. Evaporation, dehydration, freezing and packaging are more widespread in the food industry, but many major examples can be found in the chemical industry. Thus many will wonder why the food industry was slow to accept technology (and the associated disciplines) developed in the chemical industry. Two of the main reasons are hygiene and the complex nature of food.

Milk is a low viscosity fluid whose flow characteristics are much like those of water, but during processing it is vital to avoid microbiological contamination. Thus the equipment must be designed for either easy dismantling and cleaning or cleaning in place. Stainless steel is the main material of construction in the food industry, and in order to ensure that contaminants are removed, the surfaces must be smooth, free from pits and crevices and without corners. A mirror polish has often been specified, but the surface finish is less important than the roughness, which must be less than a few micrometres.

Whole milk delivered to the door is not a homogeneous product; the fat globules separate out as cream because their density is lower than that of the other components. When necessary, a homogeneous product is easily produced, and, because of its low viscosity, milk processing presents few problems. Doughs, on the other hand, are particularly difficult to handle. Firstly, they have **visco-elastic** properties resulting from a structural network of cross-linked protein. The protein is in the flour, and it cross-links when hydrated. Secondly, the adhesion characteristics, which are the bane of the inexperienced pastry cook, necessitate the treatment of all contact surfaces.

Although process control is common (for example, dough handling is often highly automated), there is more manual intervention than in the chemical industry. Figure 1.8 shows a control system consisting of a sensor (or measuring device), a comparator (which compares measured value with desired value and generates a signal for the controller), a controller and the final control element. In the food industry, the key part is the sensor — can the desired property of the material be measured? With confectionery, the coating of chocolates is a vital operation, but being a natural material, the relationships between temperature, viscosity and setting characteristics vary from batch to batch. Thus, when seeking to control the setting character-

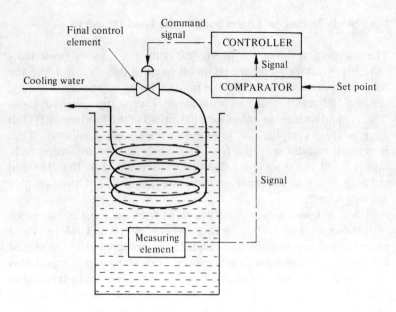

Figure 1.8 *Block diagram of simple control system. The measuring element might be a thermocouple and the final control element a control valve on a cooling water line*

istics, the measurement of temperature, which is straight-forward, will not suffice. Unfortunately, the in-line measurement of the desired properties is impossible. The result is that automatic control is supplemented by manual adjustments.

A food process engineer needs to develop a special expertise, but the same can be said of process engineers in fibre manufacture, paint manufacture, pharmaceuticals and almost all areas. Thus no chemical engineering graduate should overlook the food industry. Those graduating from departments offering chemical and bioprocess engineering courses are at a slight advantage, because subjects such as hygienic design and introductory microbiology are part of the course.

An exciting development in protein production in which chemical engineers have played a key role is protein production by fermentation. This production route will become increasingly important because of population growth, and a desire not to over-fish the sea and over-farm the land. In this relatively new route, micro-organisms are fed on simple and relatively inexpensive substances like hydrocarbons, methanol and carbohydrates. Once they have multiplied many-

fold, the cells are removed by filtering from the liquid in which they have been growing.
Table 1.4 indicates the intense interest in fermentation processes. As normal, not all of the companies which have done research and development work have considered it economical to move on to commercial production. However, some European examples are

Table 1.4 Fermentation: some of the processes that have been investigated

Feedstock	Micro-organism	Company
Methanol	Bacterium	Shell, ICI, various Japanese companies
Methane	Bacterium	Shell
Gas oil	Yeast	BP, various Japanese companies, USSR
Carbohydrate	Fungus	Rank Hovis McDougall, Du Pont, Tate & Lyle
Carbohydrate	Yeast	Swedish Sugar Corporation
Cellulose	Bacteria	Louisiana State University
Lactose	Yeast	Dairy companies
Protein content:	bacteria 75% yeasts 50% micro-fungi 45%	

ICI's methanol/**bacterium** process, Rank Hovis McDougall's carbohydrate/**fungus** process, and various **lactose/yeast** plants. The last two processes utilise industrial wastes which benefits the environment and a world of scarce resources. The first process is described in approximate terms by the following equation:

$$CH_3OH + O_2 + NH_3 \longrightarrow Cells + CO_2 + H_2O \qquad (1.9)$$

The micro-organisms grow in an aqueous medium containing methanol, ammonia and the essential nutrients. Oxygen is sparingly soluble in water and a major problem on the large scale is the dissolution of oxygen. The supply rate needs to be 2×10^{-3} g dm^{-3} s^{-1}. Also CO_2 needs to be released. Thus, ideally, in accord-

19

ance with Henry's Law which states that the concentration of a gas in solution is proportional to the partial pressure of that particular gas in the gas phase above, there needs to be an area of high pressure to promote the absorption of oxygen, and one of low pressure to promote the desorption of carbon dioxide. The preferred solution was the development of the circulating bubble column shown schematically in figure 1.9.

Although the bubble column reactor is conceptually simple, a number of questions need answering before a plant such as that shown in figure 1.10 can be built. To achieve this, production runs on a scale intermediate between laboratory scale and the full scale are undertaken. Such plants are known as pilot plants and their main purpose is to provide confirmative design data. When the product is also new, customer acceptance trials are vital, and besides using the pilot plant to provide answers to technical questions it will also be used to provide product for marketing trials.

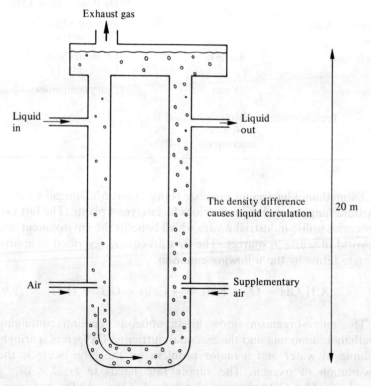

Figure 1.9 *Circulating bubble column*

Figure 1.10 ICI's protein plant. The size of the cooling tower indicates the importance of heat removal. The large size of the fermenter is also obvious

Case Study 4: Process Engineering Off-shore

Generally, process plant and equipment are designed with an exact knowledge of the raw materials, the required product and the process conditions. Usually operating experience and previous designs of similar plants are available. Off-shore oil and gas processing is a major exception. Petroleum engineers extrapolate from the evidence of a small number of exploratory wells and predict the composition and quality of oil, the pressure, the percentage of gas and the percentage of water *and* how these are anticipated to change as the oil is extracted. The process engineer may be advised that the percentage of water produced with the oil will rise from 5 per cent to 80 per cent during the production period. The quality of the oil and its pressure will also change significantly.

With all oil fields both on-shore and off-shore, this variability creates uncertainty about the feedstock. However, the off-shore environment is special, in that a typical platform may lie in 140 metres of water, 120 miles from land. The platform is massive, extremely expensive, and has been described as "the most expensive real estate in the world." The cost of a piece of equipment off-shore (including installation, operation and maintenance) is about 5–7 times that of an identical item installed on-shore. This creates the following design philosophy: keep off-shore treatment to a bare minimum, just sufficient to ensure safe transportation, and provide major treatment facilities on-shore.

The isolation of an off-shore platform necessitates that it must provide all of its own emergency services, electricity and potable water. Figure 1.11 illustrates how much is squeezed into a limited space. The value of the oil (costed at $20 per barrel) produced from the platform shown in figure 1.11 is over $100 000 per hour. This illustrates the necessity for producing particularly reliable equipment. The difficulty of maintenance (there are no off-shore workshops or machining facilities) creates the same pressure. Manpower levels are also limited by the availability of accommodation and cost. Including survival training and transportation costs, the cost of employing a man off-shore is put at £700 per day while on-shore it is less than £100 per day at 1986 prices.

As oil exploration moves into deeper water, off-shore design will differ even more from on-shore design. The major factor to date has been cost. However, future platforms will be floating structures. The wind and waves will result in heave, roll and pitching motions. The question raised for the process engineer is basically — 'will the

Figure 1.11 *A typical North Sea oil platform. Corrosion protection of the structure is essential.*

equipment cope with the inevitable sloshing?' The influence of unintentionally imposed motion on processing equipment is a relatively unexplored area of chemical engineering, but research on modified standard equipment is currently in progress. Research and development work on new types of equipment that are insensitive to imposed motion and tilt is also on-going.

Case Study Summary

Chemical engineering is concerned with the engineering of processes as well as the application of chemistry, and it is *processing* which forms the link between all of the case studies. Although these are of a very different nature, techniques of analysis and design common to all will be introduced in the rest of the book. The next chapter introduces the key integrating element of process design. Subsequent chapters look at a number of unit operations. With these tools, the process engineer selects the conditions needed to make chemical, physical or biochemical changes occur. The conditions for each unit operation will be specified and the equipment for that operation (for example, a particular type of heat exchanger) chosen. Detailed mechanical design and construction of the hardware is usually undertaken by other engineers, particularly if standard equipment is required. The development of novel processes and equipment is an inter-disciplinary task.

EXERCISES

1.1. In addition to filtration, name two other processes that can be used for solid–liquid separation.

Also name two processes that can be used for liquid–liquid separation. Do you think they have wide application throughout industry?

1.2. The block diagram in figure 1.2 does not indicate the position of the blower which is needed to drive the gases through the plant. Consider where it should be placed and record your reasons. You may wish to revise your answer later.

1.3. The laws governing chemical equilibrium can be applied to the reaction

$$SO_2 + \tfrac{1}{2}O_2 \rightleftharpoons SO_3$$

What effect will pressure have on the equilibrium position? Why do you think sulphuric acid plants are run at about atmospheric pressure? Note your answer, because this matter will be considered in a subsequent chapter.

2 Process Design

A process plant is a network of interconnected vessels joined by pipes. Obviously it is important to calculate the flows through every vessel and pipe so that they can be correctly sized. This must be done not only for the conditions which exist during normal production but also for those conditions which appertain during the start-up period and for those conditions which could occur in the event of an emergency shut-down. Certain pipes and vessels are specified solely for the latter purpose. Although it is hoped that it will not be necessary to use the emergency system, periodic testing of it (particularly the associated controls) is essential.

Knowledge of the mass and volumetric flow rates for each part of the plant results from solution of the material balance and simple cases will be examined shortly. Energy balances are also important since questions such as 'What is the thermal duty?', 'How much steam is required?' and 'Will we have sufficient cooling water?' need to be answered.

Material and energy balances thus lie at the very centre of process design. There is generally no 'right' answer, and while some of the possibilities can be eliminated on the grounds of safety or operability (ease of control), the design group is often left with a degree of choice. Final selection is based on economic considerations and company preferences.

Material and energy balances are also useful when studying plant operation. For example, the task could be to assess performance of a process such as crude oil distillation. Alternatively it might be the location of the cause of excessive material or energy loss. In subsequent sections the fundamentals are introduced together with some simple examples. A section on Process Economics is to be found at the end of the chapter.

MATERIAL BALANCES: GENERAL

The balances result simply from the application of the principle of the conservation of matter to a particular unit operation or group of processing units. It is readily understood that

$$\begin{matrix}\text{Total amount of} \\ \text{matter into vessel} \\ \text{per second}\end{matrix} = \begin{matrix}\text{Total amount of} \\ \text{matter out of vessel} \\ \text{per second}\end{matrix} + \begin{matrix}\text{Rate of accumulation} \\ \text{of total matter} \\ \text{in vessel}\end{matrix} \quad (2.1)$$

The last term is a rate or 'per second' term and would typically have units of kg s^{-1} as would the other terms.

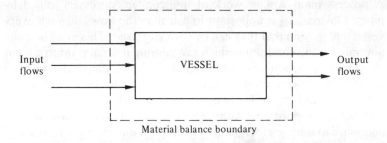

Figure 2.1 *Material balance across a vessel*

If there is no reaction, the identity of each molecule is maintained and equation (2.1) can be applied to each component in the feed streams. Hence for a general component, i, which might for example, be oxygen, ammonia, ethanol or propane:

$$\begin{matrix}\text{Amount of} \\ \text{component i} \\ \text{into vessel} \\ \text{per second}\end{matrix} = \begin{matrix}\text{Amount of} \\ \text{component i} \\ \text{out of vessel} \\ \text{per second}\end{matrix} + \begin{matrix}\text{Rate of} \\ \text{accumulation} \\ \text{of component i} \\ \text{in vessel}\end{matrix} \quad (2.2)$$

Often the accumulation term can be neglected but sometimes it is vital. To understand when it is of importance, it is necessary to understand that processes fall into two main categories: *batchwise* processes and *continuous* processes. The unit operation of **distillation** (which is covered in detail in chapter 6) can serve as a specific example. Whisky distilling is done on a relatively small scale and in a very traditional manner. A still is charged with a batch at a time and each batch is distilled consecutively. The condensed vapour varies in composition with the passage of time. There is a selective removal of the more volatile components and the concentration of these components in the liquor, from which the vapours arise, decreases with increasing time. The whisky industry traditionally produces the distillate in three distinct and successive stages. The liquor coming over early is known as foreshots, the second phase produces spirits while the last phase yields a liquor known as feints which is low in alcohol.

In an oil refinery and many parts of the chemical industry, the quantities handled are so large that it would be barely feasible and certainly uneconomic to distil one batch at a time. Thus the distillation process is run continuously. After the start-up phase has been completed, the feed rate F kg s^{-1} will equal the sum of the product rates. (Thus for figure 2.2, $F = D + B$.) Furthermore, once this

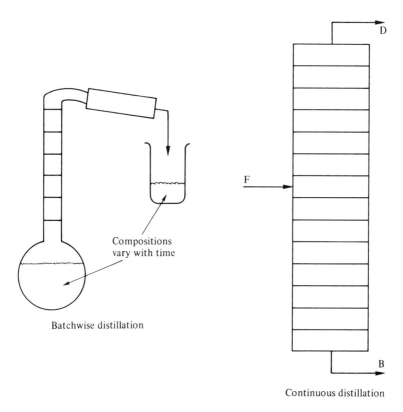

Figure 2.2 *Comparison of batchwise and continuous processes*

steady state has been achieved, the compositions at every level in the column and in the product lines will be constant provided the feed composition and rate remain fixed.

Thus for all continuous processes at steady state, equation (2.1) can be written as

$$\text{Mass flow in per unit time} = \text{Mass flow out per unit time} \quad (2.3)$$

If in addition, there is no reaction, the following holds for each component i:

$$\frac{\text{Mass flow of component i}}{\text{in per unit time}} = \frac{\text{Mass flow of component i}}{\text{out per unit time}} \quad (2.4)$$

These common-sense laws are as powerful as they are simple. The next example relates to fermentation and the following example to Case study 1.

Example 2.1

Ethanol can be produced by fermentation but the maximum concentration of ethanol in the fermentation liquid is limited to about 10 per cent (by mass) because higher concentrations have a deleterious effect upon the yeast which is essential for the reaction. Consider the concentration of 10 kg s^{-1} of a 10 per cent (by mass) aqueous ethanol solution in a continuous distillation column as shown in figure 2.2. If the distillate is 90 per cent ethanol, calculate the production rate of concentrated ethanol and the rate of water removal from the bottom of the column. Assume that the bottom product is pure water.

Solution

In addition to the above equations it is necessary to know that:

$$\frac{\text{Component flow rate in}}{\text{given stream}} = \frac{\text{Mass}}{\text{fraction}} \times \frac{\text{Total flow rate}}{\text{of that stream}}$$

Now equation (2.4) can be used for a balance related to one component and equation (2.3) for a balance on the total amounts in each stream. Balance on ethanol gives:

$$0.1 \times 10 = 0.9 \times D$$
$$\therefore D = 1.1 \text{ kg s}^{-1}$$

Total balance gives:

$$10 = D + B$$
$$\therefore B = 8.9 \text{ kg s}^{-1}$$

Thus the required answers are:

Production rate of concentrated ethanol = 1.1 kg s^{-1}
Rate of water removal = 8.9 kg s^{-1}

Example 2.2

Sulphur trioxide is removed from a gaseous stream of sulphur trioxide, nitrogen, oxygen and residual sulphur dioxide by absorption into 98 per cent sulphuric acid. The acid is maintained at 98 per cent (by mass) through continuous addition of water. If the flow rate of sulphur trioxide is 10 kg s^{-1} and 99.8 per cent of it is absorbed, calculate the daily loss of SO$_3$ in the gaseous effluent.

Solution

The flow diagram is shown in figure 2.3. Balance on SO$_3$ (stream W contains no SO$_3$):

$$10 = 9.98 + \text{mass flow of SO}_3 \text{ in stream E}$$

The mass flow rates are in units of kg s^{-1} and so daily loss of SO$_3$ is

$$(10 - 9.98) \times 3600 \times 24 = 1728 \text{ kg}$$

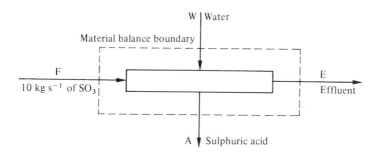

Figure 2.3 *Simple material balance example*

Material Balances: Unsteady State

Unsteady-state balances arise not only in batch processes but also when conditions in continuous processes are changed. The simple case of a two-component material balance will be considered. Consider the application of equations (2.1) and (2.2) to the vessel depicted in figure 2.4. The feed flow rate is m_F kg s^{-1}, the discharge rate m_D kg s^{-1}, the density of the liquid ρ kg m^{-3} and the volume of liquid in the vessel V m^3. Thus from equation (2.1):

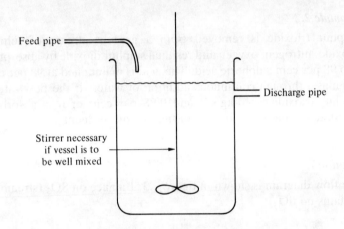

Figure 2.4 *Unsteady-state material balance on a stirred vessel*

$$m_F = m_D + \frac{d(\rho V)}{dt}$$

The last term represents the rate of accumulation of matter in the vessel. If the mass fraction of the component of interest is x_F in the feed, x_D in the discharge pipe and x in the vessel, then application of equation (2.2) gives:

$$x_F m_F = x_D m_D + \frac{d(\rho V x)}{dt}$$

Note that although x_F may be constant with time, x_D and x will vary with time. Before the above equation can be integrated, knowledge of (a) the relationship between x_D and x must be established, (b) the relationship of density to concentration ascertained, and (c) the limits on V fixed. The simplest problem results from assuming (a) the vessel to be well mixed, in which case the concentration of matter entering the discharge pipe is equal to the concentration of matter in the vessel (that is, $x_D = x$), (b) density to be constant, and (c) the vessel to be filled to the overflow level (that is, V is fixed). The above equations can then be simplified to:

$$m_F = m_D$$

$$x_F\, m_F = x\, m_F + \rho V \frac{dx}{dt}$$

The first-order differential equation is readily integrated and given an appropriate boundary condition, a solution showing how x changes with time, t, can be found. If at time $t = 0$, $x = x_0$, the solution is

$$x = x_F\,(1-\exp(-t/\bar{t})) + x_0 \exp(-t/\bar{t}) \qquad (2.5)$$

where $\bar{t} = \rho V / m_F$.

The term $\bar{t}$, which is known as the residence time, will be of use later. It is equal to the mass of liquid in the vessel divided by the mass flow rate through the vessel. Some elements of liquid in a well stirred vessel pass through in a time considerably shorter than $\bar{t}$ while others take considerably longer. Thus $\bar{t}$ represents an average time. In the following example, the concentration changes after one, two, four and eight residence times are calculated, and readers can draw their own conclusion regarding when it might be appropriate, for all practical purposes, to say that the change is complete.

Example 2.3

A tank contains 5 per cent (by mass) sodium chloride in water. It is filled to the overflow. The contents, which have a volume of 2 m^3, are well stirred and pure water is fed into the tank at a rate of 0.25 m^3 min^{-1}. Determine the concentration of salt after one, two, four and eight residence times. Calculate the respective times in minutes.

Solution

Equation (2.5) can be used, with $x_F = 0$. Assume constant density and it follows that

$$\bar{t} = 2/0.25 = 8 \text{ min}$$

Therefore, mass fraction at time $t = 0.05 \exp(-t/\bar{t})$.

Concentration (% by mass)	Time (minutes)
5.00	0
1.84	8
0.68	16
0.09	32
0.0017	64

Material Balances: With Reaction

Chemical engineering is concerned with the transformation of matter, whether it be the production of a traditional chemical, the manufacture of cement, waste-water treatment or the production of bio-mass. Balances across units in which reactions take place are therefore of great importance. Equation (2.1) is still applicable, since matter can neither be created nor destroyed except in nuclear reactions. Equation (2.2) which referred to a particular component needs to be modified because particular molecules will react or be formed by reactions. An extra 'generation' term allows for this.

$$\begin{array}{l}\text{Amount of}\\\text{component}\\\text{i into}\\\text{unit per}\\\text{second}\end{array} = \begin{array}{l}\text{Amount of}\\\text{component}\\\text{i out of}\\\text{unit per}\\\text{second}\end{array} + \begin{array}{l}\text{Rate of}\\\text{accumulation}\\\text{of component}\\\text{i in unit}\end{array} + \begin{array}{l}\text{Net rate of}\\\text{consumption}\\\text{of component}\\\text{i per second}\end{array} \quad (2.6)$$

A material balance equation can be written for every identifiable species present, and so 'component' can be understood to be either element, compound or radical. For example, in the combustion of methane, balances can be made both on the element carbon and the compound methane. When reactions are involved, it is convenient to work in terms of molar rather than mass units.

Stoichiometry is important. The stoichiometry of a particular reaction states unambiguously the ratios in which molecules of different species are consumed or formed. For example, the stoichiometry of the reaction $SO_2 + \frac{1}{2}O_2 = SO_3$ is such that the generation of one mole of SO_3 requires the consumption of one mole of SO_2 and a half a mole of oxygen.

※

$$SO_2 + \frac{1}{2}O_2 \rightarrow SO_3$$

INPUT COMP: 21 parts O_2 + 21 parts SO_2 + 79 parts $(N_2 + Ar)$
⇒ SO_2 flow = $(^{21}/_{121}) \times 500 = 86.8$ mole min^{-1} .. etc ....
SO_2 then 80% CONSUMED = $0.8 \times 86.8 = 69.4$ mole min^{-1}
O_2 ∴ 40% CONSUMED = 34.7 mole min^{-1}
SO_3 : equimolar with SO_2 ∴ 69.4 mole min^{-1} ACCUMULATES.

	(mol min^{-1})	moles (mol min^{-1})	
SO_2	86.8	−69.4*	17.4
O_2	86.8	−34.7**	52.1
N_2 + Ar	326.4	—	326.4
SO_3	—	+69.4	69.4
Total	500.0		465.3

*80 per cent of the feed SO_2 is converted.
**Amount of oxygen consumed = half amount of SO_2 consumed.

One subject of importance to chemical engineers is combustion. When a fuel is burnt in air there is a minimum of three streams: the fuel, the air and the combustion products. The nitrogen and the inert gases (argon, krypton, etc.) in the air do not react and they can be used as *tie* components to relate the inlet and outlet compositions, as shown below. The amount of oxygen is generally in excess of the stoichiometric amount to ensure complete combustion. However, too much excess air creates excessive energy losses and reduces energy efficiency. Measurement of the percentage of oxygen in the combustion products can be used to determine the percentage of excess air (per cent above stoichiometric) and the air flow is then adjusted accordingly in order to achieve the right balance.

Example 2.5

The design basis for a furnace fired with methane is 25 per cent excess air. Calculate the relative flow of air and combustion products and the composition of the flue gas on a dry basis.

The calculation will be based on 100 moles of methane, and complete combustion to carbon dioxide and water will be assumed.

Solution

$$CH_4 + 2O_2 \longrightarrow CO_2 + 2H_2O$$

Stoichiometric air: from the reaction equation, 1 mole of CH_4 requires 2 moles of O_2. Therefore inlet flows are as follows:

$$\text{amount of oxygen} = 100 \times 2(1 + 0.25) = 250 \text{ moles}$$

$$\text{amount of nitrogen} = \text{amount of oxygen} \times 79/21 = 940.5 \text{ moles}$$

Component	Amount (moles)	Change (moles)	Amount out (moles)	Composition on dry basis (%)
CH_4	100	−100	0	0.0
O_2	250	−200	50	4.6
N_2*	940.5	—	940.5	86.2
CO_2	—	+100	100	9.2
H_2O	—	+200	200	—
Total	1290.5		1290.5	

*Nitrogen term includes other inert gases such as argon.

When introducing material balances, the need to know the flows in all parts of the plant was referred to. Clearly it is also important to know the composition of the streams in the same part of the plant, and the material balances provide this information as well.

Material Balances: Recycles and Purge Streams

For a process plant consisting of a mixer, reactor and separator as shown in figure 2.5, the balances can be performed unit by unit. The product stream from the mixer is one of the feed streams to the reactor, while the product stream from the reactor is the feed to the separator. The step-by-step approach can be repeated with similar 'left to right' processing schemes, irrespective of the number of units. However, sequential calculations are impossible if a flow stream from a downstream unit is returned (recycled) to a unit upstream as shown in figure 2.6. The flow rate and composition of this recycled stream may well not be known because the calculations, for the unit from which the recycle stream came, have yet to be completed. In terms of the example in figure 2.6, the solutions for units 2 and 4 are interdependent. Simultaneous solution of the material balance equations for the units concerned becomes necessary.

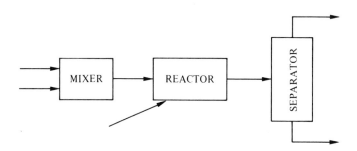

Figure 2.5 *A simple flow scheme with no recycles*

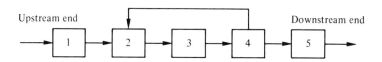

Figure 2.6 *A flow scheme with one recycle stream. Which unit is the reactor?*

Apart from simple problems, solutions are obtained by using computers. In the 1970s, these would have been large mainframes but now the task is easily handled by desk-top micros.

Recycle problems arise whenever there is a need to recycle unconverted reactant. Even near-equilibrium conversion in the

Haber process for ammonia would achieve only 25 per cent conversion in a *single pass* through the reactor, and so the nitrogen and hydrogen are recycled. The overall conversion of reactants into products can be made to be nearly 100 per cent. The synthesis of methanol from carbon monoxide and hydrogen is similar in that a recycle is essential because the conversion per pass is again around 25 per cent. The production of vinyl chloride (chloroethene) from ethylene (ethene) also involves recycle streams, and many more examples of this sort could be given. More complex processes often involve several recycle loops, and at the process design stage consideration has to be given to the effect they will have not only on steady-state operation but also on the start-up process.

The following example illustrates the benefits of recycle.

Example 2.6

Assume that the feed to an ammonia synthesis loop consists of pure nitrogen and pure hydrogen in the mole ratio of 1:3 and that the conversion per pass is limited by rate of reaction to 15 per cent. If 1 per cent of the unreacted material were to be lost with the product, show that the overall conversion exceeds 94 per cent.

Solution

The appropriate flow diagram is shown in figure 2.7. The product stream (stream 5) contains 1 per cent of the nitrogen in stream 3. The amount of nitrogen in stream 3 is 85 per cent of that in stream 2 (the rest is converted to ammonia). If the basis of the calculation is 100 moles of feed, the relationships can be symbolically represented by

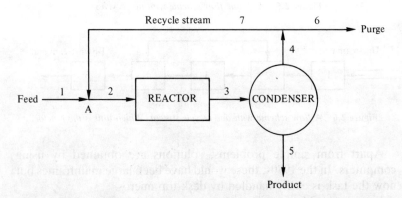

Figure 2.7 *Flow diagram for examples 2.6 and 2.7. In example 2.6, stream 6 has zero flow*

the following where, for example $F_{N_2}^4$ refers to the molar flow of nitrogen in stream 4. With no purge $F_{N_2}^6 = 0$ and $F_{N_2}^7 = F_{N_2}^4$.

Nitrogen balance at mixing point A: $25 + F_{N_2}^4 = F_{N_2}^2$

Nitrogen balance across reactor: $F_{N_2}^2 (1 - 0.15) = F_{N_2}^3$

Nitrogen balance across condenser: $F_{N_2}^3 = F_{N_2}^4 + F_{N_2}^5$

$$\text{But } F_{N_2}^5 = 0.01\, F_{N_2}^3$$

$$\therefore F_{N_2}^4 = 0.99\, F_{N_2}^3$$

From the above: $25 + (0.99 \times 0.85)F_{N_2}^2 = F_{N_2}^2$

$\therefore$ nitrogen flow in stream 2, $F_{N_2}^2 = 25/(1 - 0.99 \times 0.85)$
$\phantom{\therefore \text{nitrogen flow in stream 2, } F_{N_2}^2} = 157.7$ moles

Nitrogen loss in stream 5 = 1.34 moles

Nitrogen feed into system = 25 moles

$$\text{Overall conversion of nitrogen} = \frac{N_2 \text{ into system} - N_2 \text{ lost}}{N_2 \text{ into system}}$$

$$= \frac{25 - 1.34}{25} = 0.946$$

Thus although the conversion per pass is 15 per cent, the overall conversion of reactants into product can exceed 94 per cent.

Normally it is necessary to bleed off a fraction of the recycle stream to prevent accumulation of unwanted material. In the above example, the argon present in the feed stream was neglected. Unlike the other gases, it is not converted to product and without remedial action it would continue to build up and reach an unacceptable level. A continuous purge is normally used and if under steady-state conditions the inerts are not lost in any other stream, the basic material balance equation can be applied to the inerts to show that:

$$\text{Rate of feed of inerts into the process} = \text{Rate of loss of inerts out of process via the purge}$$

If the purge has fuel value, it can be used in the boiler house to generate steam which is a useful heat source (see chapter 4) or piped to a separator or rejected as waste. On ammonia plants the first was the preferred solution, but with the advent of membrane separators (see chapter 5) hydrogen is now recovered and recycled to the reactor.

Example 2.7

If the feed stream in example 2.6 were to contain 0.2 per cent argon, the rest being nitrogen and hydrogen in stoichiometric amounts, and the maximum desirable level of argon in the recycle stream were 4 per cent, find the purge rate as a fraction of the feed.

Solution

The flow scheme is shown in figure 2.7. Imagine a material balance boundary around the complete group of units. Thus on the basis of 100 moles of feed:

Rate of argon into system = Rate of argon removal in purge

$$\therefore 100 \times 0.2 = \text{flowrate of stream 6 } (F^6_{TOT}) \times 4$$

$$\therefore F^6_{TOT} = 5$$

∴ purge rate is 5 moles per 100 moles of feed

Material Balances: A Summary

Although the type of work outlined above will have been unfamiliar to most readers it should be clear that the flowrates and compositions of the process streams, which are obtained from the application of equations (2.1) and (2.6), are not only important but also relatively simple to obtain. Material balances in general, with reaction and with recycle have been considered. This covers the full range of possible problems and a complete process can be described by the appropriate number of equations of the same form as those introduced above. The main difference from the worked examples is simply the need to use computers to solve the simultaneous equations.

ENERGY BALANCES: GENERAL

Energy balances are essential both for **process design** and in **process analysis**. In the former case the information is required in order to size heaters, coolers and pumps. When analysing the performance of existing plant it is often necessary to compare actual energy consumption with the design figures.

Figure 2.7 does not indicate any energy requirements, but in the context of an ammonia plant the feed and recycle streams will need heating, the reactor products cooling (in order to condense the ammonia) and the gases in the reactor will also need cooling (because the exothermic reaction raises the temperature and adversely affects the equilibrium). In the absence of reaction, the material and energy balances are independent. However, if the extent of reaction is affected by temperature, the two balances must be solved simultaneously.

The energy content of a stream is conveniently represented by the specification of the enthalpy of that stream, and the units will be joules. The units of specific enthalpy are $J\ kg^{-1}$ or $J\ mol^{-1}$, and these refer to the energy content per unit amount of matter. The specific enthalpy of a substance is dependent upon temperature and pressure while that of a mixture is also dependent upon the composition.

The thermodynamic background is outside the scope of this book but an appreciation of the effect of temperature upon enthalpy is important. It also provides a link with the familiar concept of specific heat capacity.

For a pure substance with no phase change, the enthalpy h_T at a temperature T is given by:

$$h_T = \int_{T_{ref}}^{T} C_p\ dT \tag{2.7}$$

where C_p = specific heat capacity and T_{ref} = reference temperature.

If a phase transition takes place at a temperature T_p, the enthalpy h_T is given by:

$$h_T = \int_{T_{ref}}^{T_p} C_{p1}\ dT + L + \int_{T_p}^{T} C_{p2}\ dT$$

where C_{p1}, C_{p2} = specific heat capacity of phases 1 and 2 respectively, and L = latent heat of phase change (evaluated at temperature T_p).

The above equation is a logical development of equation (2.7). Both equations include a reference temperature which is often 0°C, sometimes 25°C. It is important to know the value of this reference temperature, particularly if data are being collected from more than one source. Above it was stated that enthalpy represents the energy content of a stream. It will now be appreciated that it represents the energy content of a stream, with respect to some standard state that is taken to have zero enthalpy. Today enthalpy and other relevant data of most compounds are readily found in physical property computer databases. A large amount of tabulated data is also available and table 2.1 is an illustrative extract from standard steam tables. The following example, which uses steam tables, illustrates why it is convenient to work in terms of enthalpy rather than specific heats and latent heats.

Example 2.8

Superheated steam is produced at a pressure of 5 bar and a temperature of 200°C in a vessel which is electrically heated. If the water is pumped in at 0.1 kg s^{-1} and its temperature is 20°C, determine the energy required.

Solution

energy required = enthalpy flow out − enthalpy flow in
 = water flow rate (specific enthalpy out − specific enthalpy in)

From tabulated data (which is readily found for steam) the enthalpies are 2857 kJ kg^{-1} and 83.9 kJ kg^{-1}. Thus

energy required = 0.1 (2857 − 83.9)
 = 277 kW

In the above example it was not necessary to know how the specific heats varied with temperature, at what temperature the phase change occurred and the latent heat of vaporisation at that temperature. By working in enthalpies it was sufficient to know the enthalpy of the feed stream and the enthalpy of the product stream. Thus, whenever possible, engineers extract enthalpy data from reference sources or computer databases in preference to specific heat capacity and latent heat data.

Energy Balance Equation

A stream flowing into a vessel has energy in the form of enthalpy (this term includes internal energy and pressure energy), kinetic energy and potential energy. The last two terms are often small compared with the enthalpy, heat and work terms, and it is appropriate to neglect them. Thus the energy balance for a unit is generally written as:

Table 2.1 Illustrative data from steam tables

Pressure (bar)	Saturation temperature (°C)	Specific volume of vapour ($m^3\ kg^{-1}$)	Specific enthalpy of liquid ($kJ\ kg^{-1}$)	Enthalpy due to change of phase (latent heat of vaporisation) ($kJ\ kg^{-1}$)	Specific enthalpy vapour ($kJ\ kg^{-1}$)
1.0	99.6	1.694	417	2258	2675
2.0	120.2	0.8856	505	2202	2707
5.0	151.8	0.3748	604	2109	2749
10.0	179.9	0.1944	763	2015	2798
20.0	212.4	0.0996	909	1890	2799
50.0	263.9	0.0394	1155	1639	2794
100.0	311.0	0.0180	1408	1317	2725
200.0	365.7	0.0059	1827	584	2411
221.2	374.15	0.0032	2084	0	2084

NOTE:
(a) These data amount to only 1 per cent of that available in standard steam tables.
(b) The same data are available for over 100 chemicals in most physical property computer databanks.
(c) The critical point of steam is easily identified in the above table.
(d) The standard state of water is the triple point (at which ice, water and steam can co-exist). The corresponding temperature and pressure are 0.01°C and 0.006 112 bar.

Enthalpy flow   Heat input   Enthalpy flow out
into unit   +   into unit   =   of unit per
per second   per second   second

$$ (2.8)$$

+ Work done by the fluid per second

+ Rate of accumulation of energy in unit

The heat input term allows for energy input through the walls of the vessel. For example, the fluid may be heated by steam condensing in a 'jacket' surrounding the vessel. This term (when negative) can represent the withdrawal of heat from the vessel.

Energy can exist in several forms and equation (2.8) includes a mechanical energy term to allow for addition or loss of energy in this form. The fluid can itself do work (for example, steam driving a turbine) or work can be put into the system by, for example, a pump. The other terms in the above equation parallel those in equation (2.1). For steady-state operation, equation (2.8) can be written as:

$$H_2 - H_1 = Q - W \qquad (2.9)$$

where H_2 and H_1 represent the enthalpy flows out and in respectively, Q the heat input and W the work done by the fluid.

Equation (2.9) can be applied to the production of power via the steam cycle shown diagrammatically in figure 2.8. Although this is an important area of chemical engineering, the necessary **thermodynamics** for a full analysis are beyond the scope of this book. However, the following example, which is simplified, shows that less than half of the energy available in the steam is converted into work.

Example 2.9

A steam driven turbine is used to produce work. The steam available at 180 bar and 500°C is expanded at constant **entropy** until the dryness fraction is 0.88. Calculate (a) the energy produced per kg of steam and (b) the energy produced as a fraction of the energy available in the inlet steam. How much energy in the form of latent heat is left in the exit stream?

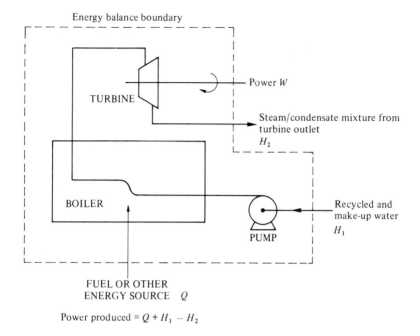

Figure 2.8 *Power production via a steam cycle — the energy Q can be supplied by burning fuel but where possible very hot process streams that need cooling (for example, the hot gases from the secondary reformer on an ammonia plant) are used*

Solution

The dryness fraction of 0.88 indicates that the stream leaving the turbine is 88 per cent steam and 12 per cent water. A constant entropy expansion is the ideal expansion for the production of maximum work, and more advanced work would show that the exit conditions are 5 bar and 152°C.

From steam tables the enthalpy of (a) the superheated inlet steam, (b) the saturated outlet vapour and (c) the saturated outlet liquid can be found. Thus:

specific enthalpy of the inlet steam = 3268 kJ kg^{-1}
specific enthalpy of the exit
steam/water mixture   = 0.88 × 2749 + 0.12 × 640
                             = 2496 kJ kg^{-1}

∴ energy produced per kg of steam = 3268 − 2496
                                  = 772 kJ

energy produced as percentage of inlet enthalpy $= \dfrac{772}{3268} \times 100$ per cent

$= 24$ per cent

latent heat energy left in product stream $= 0.88\ (2749 - 640)$

$= 1856\ \text{kJ kg}^{-1}$

This is equivalent to 57 per cent of the energy initially available in the inlet stream.

The energy source for the production of steam is not necessarily a fuel; process streams undergoing exothermic reactions are an alternative source of heat and the energy extracted from a sulphuric acid plant is as valuable, in terms of revenue, as the acid produced. Although this is the first mention of money since the previous chapter it will readily be appreciated that many process decisions, for example, the relative size of a purge stream or the choice of method for heating a vessel, must be evaluated economically as well as technically.

PROCESS ECONOMICS

This subject is an integral part of any chemical engineering course because in the last resort all process projects must be economically justifiable. Macro or national economics are not normally part of a chemical engineering course but an understanding of company finance and techniques for costing competing projects are essential. Thus, in industry there are commonly references to 'the bottom line' and 'pay-back periods'.

A distinction is often drawn between labour-intensive and capital-intensive processes. The former term is self-explanatory and alongside it the nature of the latter is also clear. Almost invariably, process plants which are costly to build and subsequently operated by relatively few staff are in the capital-intensive category. Thus with regard to a particular project a chemical company has to spend a lot of money before any product is produced for sale. Many are familiar with the long lead times of major electricity generating plants: five years for conventional oil and coal fired plants and at least seven (often ten or more) for nuclear plants. In view of the massive amounts of money which are invested in such projects it is not surprising that the *economic* case for nuclear power hinges to a large extent upon whether a modern plant can be built in seven years rather than twelve. If a person were to borrow £1000 from a bank now with the intention of paying it back in seven years time, then, given an interest rate of 15 per cent per annum, the money owed would be £2660. Retention of the loan for a further five years would increase indebtedness by 100 per cent with the outstanding amount being £5350. Although highly simplistic, the example illustrates the importance of taking into account the time factor.

The sulphuric acid plant shown in figure 1.3 took 1 year to design and 20 months to build, while the North Sea oil platform, figure 1.11, was designed in 18 months and built and installed in 2 years at a cost of about £800 million. When complex and expensive equipment is the major cost, tight financial control and thorough planning as well as a technically sound design are essential if a plant is to be built economically.

When either looking at the economics of a new process or examining the costs of an existing production plant it is useful to divide the costs into two categories. **Fixed costs** are those costs which have to be met whether or not the plant is producing. These include charges associated with:

Depreciation — related to the capital cost of the plant
Operating personnel — organised into shift teams
Maintenance staff — includes electricians, fitters and instrument technicians
Laboratory staff — analysis and quality control
Rates and insurance
Company overheads — support for finance, personnel and sales departments

Variable costs are those costs which are proportional to the rate of production. In addition to the raw materials which are converted to product, other chemicals are essential to a process. These processing chemicals may be filter aid (diatomaceous earth is used to assist the filtering of molten sulphur prior to its combustion), solvents, absorbents (although the absorbent used for carbon dioxide removal during the production of ammonia is regenerated, there are continual losses), catalysts, acids or alkalis for pH adjustment, or simply water.

The other main group of variable costs is associated with energy input and energy removal. The utilities — that is, steam, fuel oil, gas, electricity and cooling water — are the main items. Following the large rises in crude oil prices in 1973 and 1979, energy input costs became a significant variable cost, and although the crude oil price has fallen in the mid-1980s, efficient use of energy is important because the medium-term and long-term trends in the price are definitely increasing ones.

Major modern power stations built solely to produce electricity are said to be about 35 per cent efficient, because only 35 per cent of the energy available in the fuel is converted into electrical energy. Some of the energy is lost in the form of hot gases, but the main reason for the low figure is that less than half of the energy available in the steam pushed into the electricity-generating turbines can be converted into work, and hence into electricity. The steam leaving the turbine must not contain more than 12 per cent water, otherwise the turbine blades will be severely damaged. Thus nearly 90 per cent of the latent heat of the steam remains unused which is why less than half of the energy available in the steam is converted into work.

A major processing site needing both electricity and steam has always been at an advantage because the latent heat of the steam is used as process heat. The co-generation of steam and electricity gives a high overall thermal efficiency, double that of a conventional power station. The steam produced can be at various pressure levels depending on the temperature levels required; a major site might have high-pressure, medium-pressure and low-pressure steam mains. The water used has to be specially treated (mainly to prevent scaling) and so the steam condensate is returned to the boiler house for re-use because this minimises the need for fresh make-up water.

When deciding between competing designs the variation in the capital cost element may be the decisive factor, but once a process plant has been built the focus of management's attention is the variable costs. Is steam being wasted? Can the conversion of raw material to product be improved? These and similar questions will be kept under periodic review by comparing current usage of chemicals and energy with established usage.

This usage may be slightly dependent upon the production rate. While this is important when looking in detail at the efficiency of a plant, the variable costs in £ per tonne are, when painting a general picture, taken to be independent of production rate. The overall cost of production in terms of £ per tonne produced is, however, highly dependent upon production rate, because of the fixed costs. This is illustrated in figure 2.9, which shows the average cost of production falling steadily with increased production. Sales income is likely to be proportional to production, and for the hypothetical figures used in

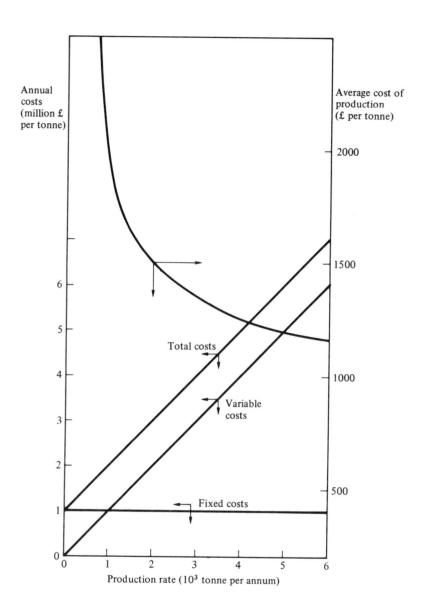

Figure 2.9 *Effect of production rate upon production costs*

figure 2.10, the breakeven point is at a production rate of 3900 tonnes per annum, that is, 65 per cent of capacity.

The profit is the difference between the sale income and the production costs and it is available for taxes, to finance growth and to

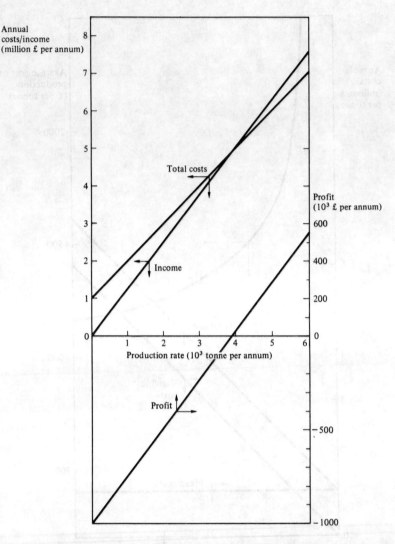

Figure 2.10 *Effect of throughput upon profitability*

pay shareholders. Profit as a percentage of the capital employed is a most important factor, and before embarking on any investment the key decision-makers would want an accurate estimate of this quantity. If money placed on deposit in a bank can earn 8 per cent per annum, no one in a free market economy is going to invest in a new production process unless the return on capital is substantially more. For production processes within an established business area, a chemical company might be looking for 15–20 per cent. When expanding into new areas, there is a higher element of risk and projects with a return of less than 30 per cent might be unacceptable.

While students of chemical engineering are taught techniques such as net present value and discounted cash flow, so as to allow for the time value of money, a full evaluation of large-scale projects will involve expertise not found within process design groups. Thus design engineers must be able to interact with those handling financial matters. This interaction and a number of others, together with the multi-disciplinary nature of process design, are illustrated in figure 2.11.

A full design of a process plant, be it a sulphuric acid plant or a food processing plant, is not within the competence of any one

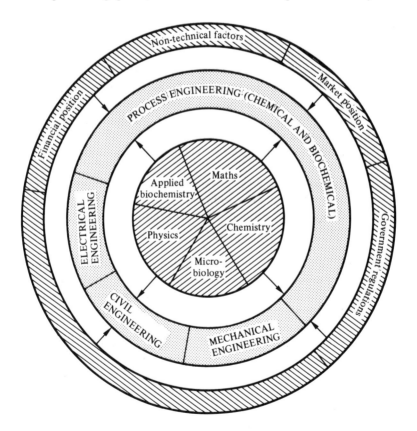

Figure 2.11 *Total Design involves a number of engineering disciplines, is influenced by a changing scientific base and is impinged upon by non-technical factors*

discipline. The evaluation of the profitability of a project involves non-technical as well as technical factors and although sales managers and economists will seek to quantify such factors as market share and the price of a barrel of crude oil in three to ten years' time, there is an element of uncertainty.

Changing economic circumstances as well as changing technology create a challenge for the chemical engineer designing or improving process plant. Within this changing picture the need for accurate mass and energy balances remains paramount, since they determine the size and design of the associated unit operations. Having introduced such balances, attention will now be given to some of the more familiar unit operations.

EXERCISES

2.1. A common question that process engineers have to answer is 'How many tonnes of X are required per tonne of product?' The quantity X can be, for example, a raw material, a catalyst or steam. In Case Study 1, process water (that is, water that mixes with the product being produced) is pumped into the SO_3 absorbers. Use the data in example 2.2 to calculate the mass flow rate in (a) units of kg s^{-1}, and (b) in tonnes per day.

2.2. List the raw materials necessary for the production of sulphuric acid and ammonia. By making appropriate use of the stoichiometric equations, estimate their consumption in kg per kg of product.

2.3. In the section on unsteady-state material balances you were asked to draw your own conclusions regarding when it might be appropriate, for all practical purposes, to say that the change in concentration was complete. Will 'for all practical purposes' be dependent upon circumstances?

A small company purchases 98 per cent sulphuric acid and produces 85 per cent sulphuric acid for internal use. The cooled mixing vessel to which water and 98 per cent acid are fed is initially filled with 98 per cent sulphuric acid. Determine the minimum time for the acid concentration to reach 86 per cent strength if the maximum water flow rate is 0.01 m^3 s^{-1} and the working volume of the vessel is 1 m^3. State any assumptions.

2.4. In the North Sea there is a weather 'window' during the summer when it is possible to tow out platforms and complete the installation. If because of delays in construction the fabrication is finished in October instead of April, the six months' delay may well cost a year's production. For a project costing £200 million generating an annual income of £50 million when 'on stream', the pay back period can be defined as capital cost/annual income, which in this case is 4 years. If the project is delayed 12 months and the interest charged on the £200 million is 25 per cent per annum, what then is the pay-back period and by how many years is the breakeven point delayed?

3 Fluid Flow

Given that a processing plant is a network of pipes and vessels, it is clearly important to be able to size every pump and all of the pipes. Thus techniques to calculate the pressure drop between the ends of each pipe are important. As detailed later, there are two main types of flow which will simply be called 'ordered **streamline**' and 'chaotic **turbulent**' for the moment. Detailed knowledge of the flow pattern is unimportant when calculating the pressure drop. Knowledge of pipe size, viscosity, density and velocity of the fluid enable the engineer to select the appropriate equation. The one equation for 'ordered streamline' flow and the small set of equations for 'chaotic turbulent' flow will be given later.

Fluid flow affects the performance of numerous pieces of process equipment and some of the examples mentioned in subsequent chapters are heat transfer, distillation, gas **absorption** and **membrane filtration**.

NEWTONIAN AND NON-NEWTONIAN FLUIDS

Viscosity is a measure of fluid friction and is proportional to the force required to move a layer of fluid over another layer. Highly viscous materials are those that possess a great deal of internal friction when layers are in relative motion — they cannot be spread or poured as easily as less viscous materials. Consider figure 3.1 in which two parallel planes of fluid of equal area A, separated by a distance d, are moving in the same direction but at different speeds. The force F which is necessary to maintain this difference is given, for many fluids, by the following equation which was introduced by Sir Isaac Newton:

$$\frac{F}{A} = \mu \frac{\Delta v}{d} \qquad (3.1)$$

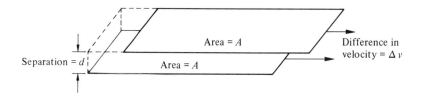

Figure 3.1 *Model for defining viscosity*

where Δv is the difference in velocity and μ is a constant for a given material.

The velocity gradient $\Delta v/d$ is a measure of the rate with which velocity changes with distance and it measures the shearing that the fluid experiences. The term is often called the 'rate of shear'. The force per unit area that is required for the maintenance of the shearing action, F/A, is known as the shear stress. A 'stress' is a force per unit area and has the same units as pressure.

At a given temperature, viscosity is, for many fluids, independent of the rate of shear, and if $\Delta v/d$ is doubled then the shear stress F/A is also doubled. Fluids of this type are known as 'Newtonian' and are best exemplified by gases and liquids such as water, petrol, thin motor oils and milk.

Figure 3.2 shows that as milk is concentrated, the nature of the fluid changes. At low concentrations it behaves as a Newtonian fluid with viscosity being constant for all rates of shear. However, above 25 per cent protein, the material is particularly viscous at low rates of shear. This fact is very important in cheese manufacture and those designing dairy plants have to ensure that a minimum rate of shear is maintained in the processing equipment which will include heat exchangers and maybe membrane units. The above behaviour, known as **shear-thinning** or pseudoplasticity is typical of polymer solutions and liquids with a second phase in suspension. Their behaviour can often be represented by an equation of the form:

$$\frac{F}{A} = C\left(\frac{\Delta v}{d}\right)^n \qquad (3.2)$$

where C is a consistency constant and n is less than unity.

Curves A and B of figure 3.3 illustrate visco-plastic behaviour. Fluids of this type will not flow until a critical **yield stress** is exceeded. Concentrated tomato ketchup is a good example. Although less

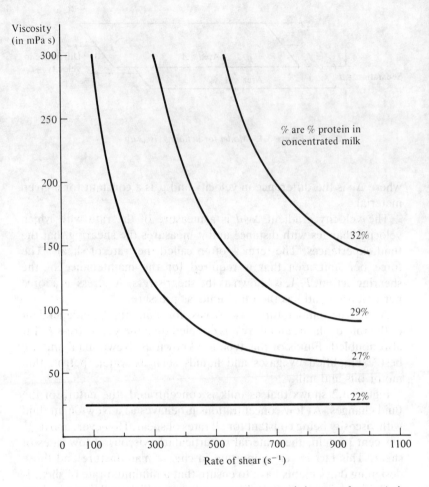

Figure 3.2 *Dependency of viscosity upon concentration and shear rate for a typical shear-thinning liquid*

obvious with modern formulations, it is sometimes difficult to pour ketchup from a bottle. However, when the bottle is shaken or struck, the yield stress is exceeded and the ketchup starts to flow, often in a bit of a gush.

There is a further group of fluids which show time-dependent behaviour. The viscosity of these materials is affected by the amount of shearing applied in the recent past to the material. Often the viscosity will decrease with time during shear but recover, sometimes slowly, when the shear stress is removed. This particular behaviour is termed thixotropic and some paints have this property.

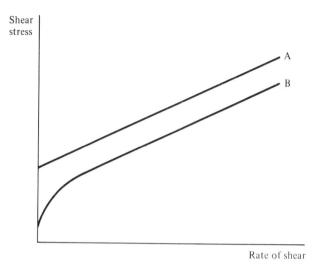

Figure 3.3 *Relationship between shear stress and rate of shear for visco-plastic fluids*

If a fluid subjected to shear suffers an irreversible decrease in viscosity then shear breakdown or rheodestruction is said to have occurred.

Clearly, the flow properties of fluids being processed are very important. Non-Newtonian behaviour is exhibited by a wide range of industrially important liquids and knowledge of this subject is and will be of increasing importance, not least because many bioprocessing liquids are non-Newtonian.

NATURE OF FLUID FLOW

Consider the apparatus shown in figure 3.4. Will the dyed water that is introduced via the needle form a coloured filament as shown in figure 3.5(a) or will it be dispersed across the whole cross-section of the pipe as shown in figure 3.5(b)? The former corresponds to the ordered streamline state mentioned earlier and is characterised by the absence of convective movement in the radial direction. Thus every fluid element reaches, once any initial disturbances have been dampened, a constant velocity that is parallel to the pipe axis.

At the pipe wall it is known that there is no slip (that is, the molecules adjacent to the solid surface are stationary) and so the

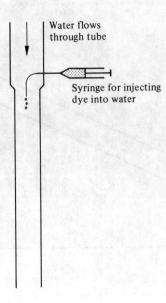

Figure 3.4 *Schematic diagram of apparatus for investigating the nature of fluid flow*

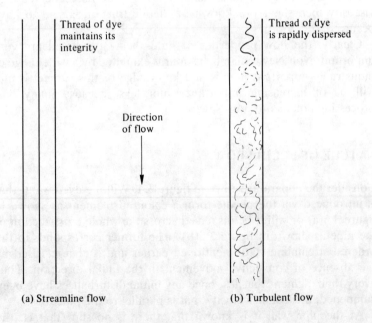

Figure 3.5 *Illustration of the two main types of flow*

velocity at the wall is zero. Given this boundary condition, the solution of the appropriate equation shows that there is a parabolic variation of velocity u with radial position:

$$\frac{u}{u_{av}} = 2\left(1 - \frac{r^2}{(d/2)^2}\right) \quad (3.3)$$

where u_{av} = volumetric flow rate/cross-section area
r = radial co-ordinate measured from the axis
d = pipe diameter.

If the above flow is disturbed by addition of a roughened section of pipe or a valve, it will either revert to the ordered state of **streamline** or **laminar** flow, or the induced oscillations become stable and a high degree of radial mixing occurs. In this turbulent state a rapid random motion is imposed on top of the time averaged motion. The latter is given by the following equation and is illustrated in figure 3.6.

$$\frac{u}{u_{av}} = 1.22\left(1 - \frac{r}{d/2}\right)^{1/7} \quad (3.4)$$

The turbulent velocity profile is flatter than that for streamline flow and the maximum velocity is only 22 per cent greater than the average velocity. However, the main difference is the rapid random motion which rapidly mixes the fluid elements. If the pipe is part of a heat exchanger in which the fluid in the pipe is being heated by, for example, steam condensing on the outside of the pipe, then the mixing process exchanges hot fluid elements from the wall area with cold fluid elements from the central area. Thus the fluid rapidly receives heat from the condensing steam. The corollary of this desirable state of affairs is that the designer of process equipment often seeks to induce turbulent conditions.

The flow pattern varies with fluid velocity, u, density, ρ, and viscosity, μ, of the fluid and the geometry of the system. For flow in a pipe the characteristic dimension is the diameter d and the viscous forces can be taken to be proportional to $\mu u/d$, which is of the same form as equation (3.1). By assuming that the flow pattern is determined by the *ratio* of inertial forces (which are proportional to ρu^2) to viscous forces, it can be shown that the key parameter is $\rho u d/\mu$. This is dimensionless and is known as the **Reynolds number** after a famous nineteenth century experimentalist who established that this

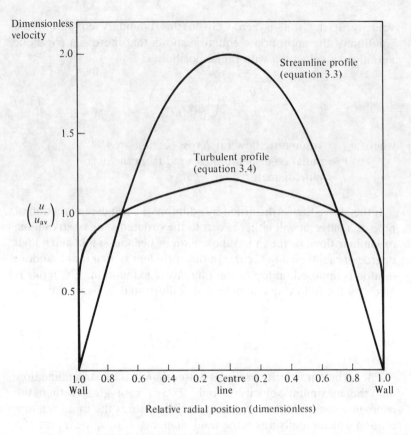

Figure 3.6 *Laminar and turbulent velocity profiles*

was the key criterion for flow in pipes. Below a Reynolds number of 2000 the flow is streamline. Above a value of 3000 it is turbulent.

With regard to pipe flow it is irrelevant whether the pipe is vertical or horizontal, and gravitational forces do not affect the nature of the flow pattern. In mixing vessels on the other hand, gravitational, inertial and viscous forces can all be important and the value of the gravitational constant, g, has to be taken into account. The resulting analysis gives rise to two dimensionless ratios: the Reynolds number introduced above and a second dimensionless number which represents the ratio of inertial to gravitational forces. More dimensionless groups will be met later on and they are favoured by engineers and mathematicians precisely because they are dimensionless and this greatly helps in the formulation of general relationships. For example, the values of 2000 and 3000 mentioned above apply to all fluids and all pipe sizes and one would confidently expect the same result in a zero gravity environment.

PRESSURE DROP IN A PIPE

In all of our Case studies and in chemical engineering generally, pressure drop in pipe lines and across items of equipment is important. Power has to be supplied to overcome the viscous forces at the walls of the pipes. Thus the key element is the shear stress at the wall τ_0. The retarding viscous force is $\tau_0 \pi dL$, where L is the pipe length, and this is balanced by the pressure driving force $\Delta p \pi d^2/4$, as illustrated in figure 3.7. At steady state these forces balance and rearrangement gives:

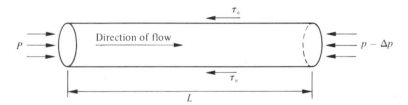

Figure 3.7 *Illustration of key parameters in force balance*

$$\Delta p = 4\frac{L}{d}\tau_0 \qquad (3.5)$$

For streamline flow, equation (3.3) applies right up to the wall and the velocity gradient at the wall can be readily calculated, and for a Newtonian fluid the shear stress obtained. It is, as those who attempt exercise **3.1** will find:

$$\tau_0 = \mu 8 u_{av}/d \qquad (3.6)$$

Combining equations (3.5) and (3.6), the pressure drop in a long horizontal pipe with streamline flow is shown to be:

$$\Delta p = 32\mu u_{av} L/d^2 \qquad (3.7)$$

Equations (3.4) and (3.5) cannot be combined and a corresponding equation obtained for turbulent flow because (a) the turbulent eddies, which are not part of a time-averaged equation, do contribute to the shear stress, and (b) equation (3.4) does not apply in the wall region. At the wall itself, the velocity is zero and close to it the velocity is obviously small. Thus, the local conditions are in fact laminar and turbulent flow consists of a large turbulent core and a small laminar sub-layer at the boundary. The picture is further complicated by the roughness of the wall. If the roughness size, e is less than the laminar sub-layer, the wall is hydraulically smooth. However, for most commercial pipes the roughness is important and the pressure drop is a function of both the Reynolds number and the relative roughness e/d. This somewhat advanced subject matter is developed in the exercises.

The change in pressure drop per unit length with velocity is of interest. For streamline flow the two quantities are proportional, as indicated by equation (3.7). Beyond a Reynolds number of 3000 the pressure gradient is proportional to the velocity raised to the power of 1.8. Thus in turbulent flow the pressure drop increases much more rapidly with velocity than it does in streamline flow, and this is one of the few real disadvantages of turbulent conditions.

In the next chapter it will be seen that the amount of heat transfer under a given set of conditions is strongly influenced by the Reynolds number. This is equivalent to stating that heat transfer is strongly influenced by fluid flow. Links between fluid flow and mass transfer will emerge in chapters 5, 6 and 7.

EXERCISES

3.1. By differentiating equation (3.3) and combining it with equation (3.1) show that the shear stress at the wall of a pipe is correctly given by equation (3.6).

What is the shear stress at the centre of the pipe?

3.2. Calculate the Reynolds number for each of the following flows and so determine whether the flow is streamline or turbulent. Four fluids (air, carbon dioxide, water and sulphuric acid) are to be considered for the flowrates given and for both a 15 mm pipe and a 150 mm pipe.

	Air	Carbon dioxide	Water	Sulphuric acid
Flow 1($m^3\ s^{-1}$)	0.001	0.001	50×10^{-6}	50×10^{-6}
Flow 2($m^3\ s^{-1}$)	0.01	0.01	0.001	0.001
Density (kg m^{-3})	1.2	1.8	998	1800
Viscosity (N s m^{-2})	18×10^{-6}	15×10^{-6}	0.001	0.01

3.3. For two cases of streamline flow (selected from those identified in the above example) calculate the pressure drop per 100 metres of pipe.

3.4. Select two examples of turbulent flow from exercise **3.2** and use the following approximate equation, which is applicable to smooth uncorroded pipes, to calculate the pressure drop for a pipe 100 m long:

$$\Delta p = 2C_\mathrm{f} \rho u_\mathrm{av}^2 L/d \text{ where } C_\mathrm{f} = 0.079 Re^{-0.25}$$

The dimensionless term C_f is known as the friction factor.

3.5. Repeat exercise **3.4** using the following equation which is one of several available for rough pipes:

$$\Delta p = 2C_\mathrm{f} \rho u_\mathrm{av}^2 L/d$$

where

$$C_\mathrm{f} = 0.001\ 375(1+(20\ 000 e/d + 10^6/Re)^{0.33})$$

Typical values of e/d range from 0.0001 to 0.01, and the effect of this ratio can be investigated. The equation is said to be accurate to within ±5 per cent for Reynolds numbers between 4000 and 10^7.

4 Heat Transfer

The transfer of heat from one fluid to another is of crucial importance but is just one of a number of transfer processes. Before looking at one particular aspect of a complex whole a few of the inter-relationships will be mentioned.

The three subjects of heat, mass and momentum transfer are inter-linked; heat flows down a temperature gradient, mass flows down a concentration gradient and momentum passes down a velocity gradient. These inter-relationships, which are firmly based on fundamental equations, are useful. In the previous chapter some introductory aspects of fluid flow were examined but it is important to realise that the three subjects mentioned above are all inter-dependent and an umbrella term 'transport phenomena' is often used to group them together.

On a chemical engineering course the subject of fluid flow (which at a fundamental level is related to momentum transfer) will include not only the flow of gases and the flow of liquids but also two-phase flows such as fluid–solid and gas–liquid flows. An understanding of gas–liquid flow is important if the operation of boilers (and every distillation column has a boiler to provide the upward-flowing vapour) is to be understood. Unfortunately, the subject of two-phase flow though very interesting, is too complex for this introductory text.

In the food industry the inter-relationship between micro-organism growth, temperature and time is of vital importance. For the destruction of micro-organisms and the inactivation of **enzymes**, heat treatment may well be selected. Other possibilities include chemicals and ionising radiation.

The importance of heat transfer to chemical engineering will be illustrated by returning briefly to Case study 1. In the sulphuric acid process there are three vital heat transfer operations:

(a) removal of heat from the SO_2 rich inlet stream

(b) removal of heat from the reactor, and
(c) removal of the heat of absorption.

In the first two cases, steam is raised by using the high-grade (high-temperature) heat to evaporate water in a boiler. This simultaneously cools the gas stream to reactor temperature and allows the heat to be used elsewhere. This last statement may seem surprising, but steam is an excellent carrier of heat; for every 1 kg condensing, about 2000 kJ of heat (the exact value is pressure dependent) are released. Thus a steam flow of 1 kg s^{-1} is equivalent to a useful energy flow of over 2.0 MW. Not only is a unit mass of steam able to store and carry a large quantity of heat, but it does so at a temperature at which it can conveniently be used. It is also worth stating the obvious — it is generated from water, which is harmless and readily available.

Hardware suitable for a variety of heat transfer operations will be considered later. First, some key concepts will be introduced by way of a familiar example. The starting point will be the well understood subject of thermal conductivity.

THERMAL CONDUCTIVITY

The key question when sizing a heat exchanger is 'if X watts of heat are transferred from the first stream to the second, how large is the heat exchanger?' Knowledge of how much heat can flow through a unit area under a given temperature difference and a method for calculating that temperature difference are required. The latter will be discussed later. The former quantity echoes the definition of thermal conductivity, k.

If the quantity of heat passing per unit time through a thin slab of thickness t, and uniform cross-sectional area A, across which there is a small temperature difference $(\theta_1 - \theta_2)$, is q, then the thermal conductivity of the material, k is defined by the equation:

$$q = kA(\theta_1 - \theta_2)/t \qquad (4.1)$$

Energy is transferred from the hotter to the cooler body, and so heat is said to flow towards the cooler one. Although energy is a scalar quantity, the flux of energy (energy flow through a surface) is a vector — direction matters. The above equation is thus more precisely written in calculus notation as:

$$q = -kA(d\theta/dx) \qquad (4.2)$$

The positive direction is, as shown in figure 4.1, taken as the direction of heat flow, and so as x increases, θ decreases, which makes the temperature gradient, $d\theta/dx$, negative. (It can be said that heat flows down a temperature gradient.) Inserting the negative sign makes k positive.

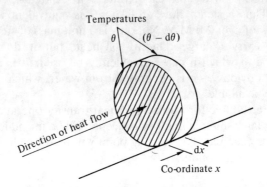

Figure 4.1 *Flow of heat through thin slab*

Some measured values of the thermal conductivity of common materials are given in table 4.1 which shows that while the temperature dependence is slight, there is a wide variation between metals (which have delocalised electrons) and the non-metals, where heat transport is by the transfer of vibrational energy from one neighbouring atom to the next.

The calculation of heat flow along a lagged bar, as shown in figure 4.2, involves a straight-forward application of the above equation. The composite slab problem is more interesting. Suppose we have two slabs of equal area A, thicknesses t_1 and t_2, and thermal conductivities k_1 and k_2 respectively. Let the temperatures be defined by figure 4.3.

Now the flow of heat through each slab is the same, therefore:

$$q = k_1 A(\theta_{hot} - \theta_i)/t_1 = k_2 A(\theta_i - \theta_{cold})/t_2 \qquad (4.3)$$

The interfacial temperature will rarely be known, but there are two equations and q and θ_1 are generally the unknowns.

Table 4.1 Thermal conductivity of common materials

Materials	Thermal conductivity k ($W\ m^{-1}\ K^{-1}$) at			
	200 K	300 K	400 K	500 K
Certain solids				
Copper (pure)	413	394	392	388
Copper (commercial)	—	372	—	—
Iron	94	80	69	61
Nylon	0.28	0.30	—	—
Titanium	26	21	20	20
Insulation materials				
Asbestos fibre	—	0.088	0.11	0.12
Mineral wool	—	0.05	—	—
Liquids				
Water	—	0.60		
Toluene	—	0.14	—	—
Mercury	—.	8.0	—	—
Gases				
Air	—	0.024	—	—
Steam	—	0.025	—	—
Methane	—	0.030	—	—

Rearrangement gives

$$\theta_{hot} - \theta_i = \frac{qt_1}{k_1 A}$$

$$\theta_i - \theta_{cold} = \frac{qt_2}{k_2 A}$$

(4.4)

Thus

$$\theta_{hot} - \theta_{cold} = q\left(\frac{t_1}{k_1 A} + \frac{t_2}{k_2 A}\right) \quad (4.5)$$

that is

$$q = (\theta_{hot} - \theta_{cold}) \bigg/ \left(\frac{t_1}{k_1 A} + \frac{t_2}{k_2 A}\right) = (\theta_{hot} - \theta_{cold})A \bigg/ \left(\frac{t_1}{k_1} + \frac{t_2}{k_2}\right)$$

(4.6)

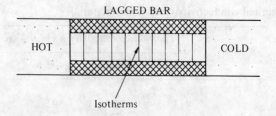

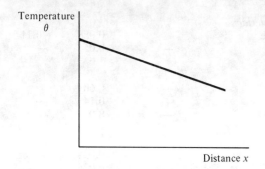

Figure 4.2 *Flow of heat along a lagged bar*

This is similar to Ohm's law; q is the flow of heat instead of current; $\theta_{hot} - \theta_{cold}$ is equivalent to the driving force, the potential difference; and the t/kA terms are thermal resistances. The equation can be generalised to give the heat flow through a composite of many layers:

$$q = (\theta_{hot} - \theta_{cold})A \bigg/ \left(\frac{t_1}{k_1} + \frac{t_2}{k_2} + \frac{t_3}{k_3} + \ldots\right) \qquad (4.7)$$

where θ_{hot} and θ_{cold} are the temperatures of the outer surfaces of the composite.

Heat Loss Across Windows: An Oversimplification

Our familiar example will be heat loss through closed windows. An attempt will be made to estimate the loss using the above theory. Single and double glazed windows of the following specifications will be assumed. Single glazed 4 mm thick glass with $k = 1.05$ W m^{-1} K^{-1}.

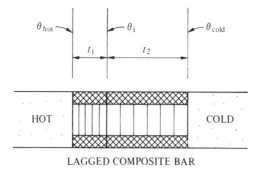

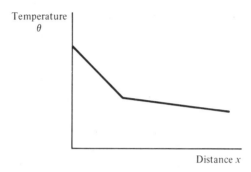

Note: Large negative $d\theta/dx$ implies small k

Figure 4.3 *Definition of temperatures for composite slab*

Double glazed units incorporating two panes of 4 mm thick glass and a 12 mm air gap, whose thermal conductivity is taken to be 0.023 W m^{-1} K^{-1}. The area of glazing will be taken as 3 m^2, room temperature as 20°C and outside air temperature as −4°C.

It could be argued that for the calculations the temperatures should be in Kelvin, not degrees Celsius. However, the numerical results are not affected, since temperature *differences* are the same in K and °C. Normal engineering practice does not slavishly follow the SI set of units and °C will be retained. Application of equation (4.7) leads to the following estimates:

$$\text{Heat loss through double glazing} = [20 - (-4)]3 \bigg/ \left(\frac{0.004}{1.05} + \frac{0.012}{0.023} + \frac{0.004}{1.05}\right)$$

$$= 136 \text{ W}$$

67

Heat loss through
single glazing: $= [20 - (-4)]3 \Big/ \left(\dfrac{0.004}{1.05}\right)$

$= 18\,900$ W

The last figure is clearly excessive, since 18.9 kW is greater than the heat input for a whole house! If the inside surface of the pane were 20°C and the outer −4°C, then the heat loss would undoubtedly be in excess of 18 kW, but are the temperature gradients shown in figure 4.4(a) reasonable? It is important to be explicit about one's assumptions. Figure 4.4(a) implies that the outside air close to, and right up to, the window is all at −4°C, despite a large outflow of heat. Similarly, there is no temperature gradient on the room side. The model which was implicitly assumed, and which has been made explicit in figure 4.4(a), is unrealistic. The model illustrated in figure 4.4(b) is much more realistic, but still not exact. An engineer learns the importance of using intelligent approximations and of making the best possible estimate from incomplete information, and in a small way this is illustrated by the current problem.

Having intuitively noted that there are regions close to both glass–air boundaries over which the temperature changes from bulk air temperature to glass temperature, a physical picture is required so that calculations can be performed. A reasonable approximation is to assume that the air, both on the inside and the outside, can be represented by a near stagnant **boundary layer** across which there is

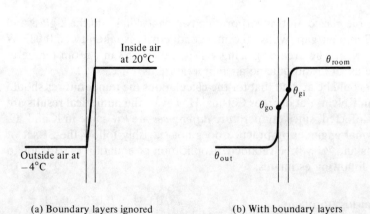

Figure 4.4 *Temperature profiles across a pane of glass: (a) ignoring boundary layers; (b) with boundary layers*

an appreciable temperature change, and a well mixed bulk which is isothermal. Reasonable thicknesses would be 2 mm for the room side, and 1.5 mm for the outside, if the wind speed is low, and 1 mm for the outside if the wind speed is higher. Thus recalculating:

Heat loss through
single glazing $= [20 - (-4)]3 \Big/ \left(\dfrac{0.002}{0.023} + \dfrac{0.004}{1.05} + \dfrac{0.0015}{0.023}\right)$
(low wind speed)

$= 462$ W

Heat loss through
single glazing $= [20 - (-4)]3 \Big/ \left(\dfrac{0.002}{0.023} + \dfrac{0.004}{1.05} + \dfrac{0.001}{0.023}\right)$
(high wind speed)

$= 536$ W

The thicknesses and the resulting heat loss values are reasonable and the model (which is one of pure conduction through a stagnant layer) might be of interest and in some circumstances of use. However, the main requirement is to have a value for the thermal resistance and it does not matter if the heat loss mechanism is a combination of convection and conduction, provided an accurate estimate can be made. In the above example, the inside thermal resistance $(t/(kA))$ is $0.002/(0.023 \times 3) = 0.029$ K W^{-1}. The reciprocal (kA/t) is normally used when referring to fluids and it is converted into a per area form (k/t). It is called the *heat transfer coefficient* and in this case would equal $0.023/0.002 = 11.5$ W m^{-2}K^{-1}. Although insights into the physics underpinning heat transfer coefficients lead to a better understanding, the coefficients are not tied to any particular model and can be treated as purely empirical constants of proportionality, the knowledge of which permits calculation of heat loss given surface area and temperature difference. The coefficients are of engineering importance and appropriate methods for estimating them are found in chemical engineering handbooks. An example is given later in the section 'Heat transfer coefficients, U-values and fouling factors'.

When heat is transferred from one fluid through a solid boundary to another, fluid heat transfer coefficients are derived for the boundary layers on each side. The method for combining these coefficients is similar to the method for combining thermal resistances and an analogue of equation (4.7) will be obtained. The temperatures for the current example are defined in figure 4.4(b), remembering that the heat flow through the glass and both boundary layers or films is the same:

$$q = h_{room}A(\theta_{room} - \theta_{gi}) = k_g A(\theta_{gi} - \theta_{go})/t_g = h_{out}A(\theta_{go} - \theta_{out}) \quad (4.8)$$

where h_{room} is the heat transfer coefficient for the inside (or room-side) boundary layer and h_{out} is the heat transfer coefficient for the outside boundary layer.

Rearrangement and addition as before gives:

$$\theta_{room} - \theta_{out} = \frac{q}{A}\left(\frac{1}{h_{room}} + \frac{t_g}{k_g} + \frac{1}{h_{out}}\right)$$

or (4.9)

$$q = (\theta_{room} - \theta_{out})A \bigg/ \left(\frac{1}{h_{room}} + \frac{t_g}{k_g} + \frac{1}{h_{out}}\right)$$

The outside heat transfer coefficient will be dependent on wind speed and window position, which need to be determined, but the exact mode of heat transport (for example, the balance between convection and conduction) is unimportant and of scientific, not engineering, interest.

The rate of flow of heat per unit area per unit temperature difference is

$$\frac{q}{A(\theta_{room} - \theta_{out})}$$

and is called the overall heat transfer coefficient, U. From the above equation, its relationship to the individual coefficients is seen to be of a reciprocal nature:

$$\frac{1}{U} = \frac{1}{h_{room}} + \frac{t_g}{k_g} + \frac{1}{h_{out}} \quad (4.10)$$

The term on the left-hand side is the overall resistance to heat transfer and those on the right are the individual resistances. This is analogous to the summing of electrical resistances.

In concluding this section, it is noted that the assumption of a stagnant layer of air between the two pairs of the double glazed units was also an oversimplification. The circulation currents within the enclosed space reduce the insulating effect. In order to reduce this loss of insulating power, certain manufacturers fill the space with inert gases which are several times denser than air. Frame construction also influences heat loss and the final overall heat transfer coefficients range from 2.0 to 3.5 W m^{-2} K^{-1} for double glazed units. This compares favourably with the 7 W m^{-2} K^{-1} of typical single glazed windows, but is not as dramatically different as many first think.

RADIAL HEAT FLOW

A common form of heat exchanger is the shell and tube type in which heat flows radially through the tube wall. A schematic diagram is shown in figure 4.5. The equivalent equation to equation (4.2) is that given below. The temperature gradient is now $\pm d\theta/dr$, the sign being

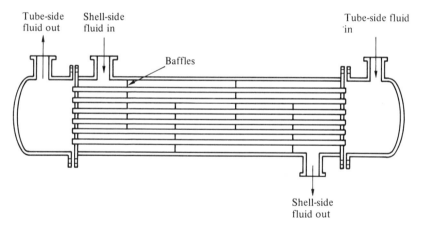

Figure 4.5 *Schematic diagram of typical shell and tube heat exchanger. Why are baffles used? What effect do they have on the shell-side flow?*

negative if the tube-side fluid is being cooled (that is, heat flows outwards). Taking a tube of length L and radius r, the heat flow through the tube wall of thermal conductivity k_w is

$$q = \pm k_w \times 2\pi rL \times d\theta/dr \qquad (4.11)$$

The development of the equivalent of equations (4.9) and (4.10) is more involved, because the area available for heat transfer depends upon the radius (that is, per unit length, the inside surface area of a tube is less than the outside surface area). This factor is trivial for thin-walled tubes, but, as illustrated in figure 4.6, is significant for thick-walled ones and lagged pipes.

The following parallels the earlier derivation, and figure 4.7 is related to figure 4.4(b)*. Consider a pipe of internal radius r_{in} and external radius r_o containing fluid at temperature θ_A. Let the shell-side fluid temperature be θ_B and the interfacial temperature at the tube-side fluid/metal and metal/shell-side fluid interfaces be θ_1 and θ_2 respectively. Then at a general radial position, r, within the pipe wall, equation (4.11) is applied, together with the relevant boundary conditions (at $r = r_{in}$, $\theta = \theta_1$; at $r = r_o$, $\theta = \theta_2$) and on rearrangement it becomes:

$$q \int_{r_{in}}^{r_o} \frac{dr}{r} = -2\pi k_w L \int_{\theta_1}^{\theta_2} d\theta \qquad (4.12)$$

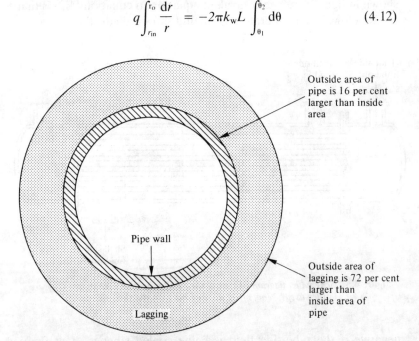

Figure 4.6 *Schematic diagram of a lagged pipe showing the significant variation in heat transfer areas per unit length*

*The reader who has difficulty following the arguments should move to the end of this section and simply compare equation (4.15) with equation (4.10).

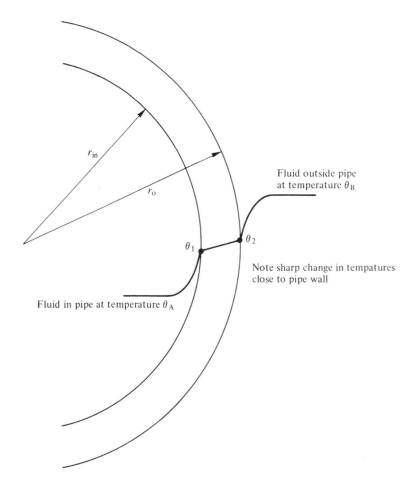

Figure 4.7 *Heat transfer across a curved boundary*

Integration and rearrangement shows that the heat flux through the pipe wall is

$$q = [2\pi L k_w / \ln(r_o/r_{in})](\theta_1 - \theta_2)$$

The other heat flux equations are

$$q = h_{in} A_{in}(\theta_A - \theta_1) = 2\pi r_{in} L h_{in}(\theta_A - \theta_1) \quad (4.13)$$

and

$$q = h_o A_o(\theta_2 - \theta_B) = 2\pi r_o L h_o(\theta_2 - \theta_B)$$

where h_{in} and h_o are the inside and outside fluid film heat transfer coefficients respectively.

Manipulation of these three heat flux equations gives:

$$\frac{q}{2\pi L}\left(\frac{1}{r_{in}h_{in}} + \frac{\ln(r_o/r_{in})}{k_w} + \frac{1}{r_o h_o}\right) = \theta_A - \theta_B \qquad (4.14)$$

Now by definition, $q = UA(\theta_A - \theta_B)$ and so the value of U, the overall heat transfer coefficient, will depend upon the value of A chosen. This does not matter, *provided that* the area to which U is referred is stated. There are two possibilities in this case and we will use $A = 2\pi r_{in} L$. Thus:

$$\frac{1}{U_{in}} = \frac{1}{h_{in}} + \frac{r_{in}\ln(r_o/r_{in})}{k_w} + \frac{r_{in}}{r_o h_o} \qquad (4.15)$$

The subscript on U denotes that we are referring U to the inner area. The first stage in any design calculations involving a radial co-ordinate system is to base the coefficients upon either the inner or outer surface *and* to state clearly the chosen reference point.

Heat Transfer Coefficients, *U*-values and Fouling Factors

The shell and tube heat exchanger shown in figure 4.5 is a very common form of heat exchanger, and the analysis of its performance involves two film coefficients. The magnitude of the individual fluid film coefficients depends on the physical properties of the fluids, the fluid flow rates, and the size and shape of the heat transfer surface. On the shell side, the flow pattern is complex and the actual velocity of the fluid across the pipes is not well defined. It is dependent on a number of variables, including pipe size, layout and spacing. However, for flow inside pipes, the geometry is well defined and the fluid film coefficient is easily estimated from dimensionless correlations derived from experimental data. The following equation is often used for fluids of moderate and low viscosity:

$$Nu = 0.023\, Re^{0.8} Pr^{0.4} \qquad (4.16)$$

where Nu = Nusselt number = $h_i d/k_f$
Re = Reynolds number = $\rho u d/\mu$ = $4G/(\pi d \mu)$
Pr = Prandtl number = $\mu C_p/k_f$

and

C_p = fluid specific heat capacity (J kg^{-1} K^{-1})
d = inside pipe diameter (m)
G = mass flowrate (kg s^{-1})
h_i = inside fluid film coefficient (W m^{-2} K^{-1})
k_f = fluid thermal conductivity (W m^{-1} K^{-1})
u = fluid velocity (m s^{-1})
ρ = fluid density (kg m^{-3})
μ = fluid viscosity at bulk fluid temperature (N s m^{-2}).

It is readily appreciated that all the quantities on the right-hand side are likely to be known, thus allowing ready calculation of Nu and hence h_i. The Reynolds number is a measure of the degree of turbulence and the equation is applicable for $Re > 10\,000$ at which turbulent flow is fully established. Prandtl number is solely dependent on the thermophysical properties of the fluid. A different equation has been developed for laminar flow. It includes a dependency on pipe length as well as diameter.

Table 4.2 shows the wide variation in values. For example, gases on the shell side (that is, on the outside of the tubes) have h values which are over an order of magnitude less than those for liquids flowing inside pipes.

Now equation (4.15) shows that the overall thermal resistance (1/U) is dominated by the largest individual thermal resistance (that

Table 4.2 Individual film heat transfer coefficients

Heat transfer operation	Usual range for h(W m^{-2} K^{-1})
Gas flow across tube banks	10– 50
Liquid flow across tube banks	100– 1000
Organic liquid inside tubes	250– 2500
Water flowing inside tubes	500– 5000
Boiling liquids	1000–10000
Condensing vapours	1500–25000

is, the smallest value of h) just as the overall resistance in an electrical circuit, containing resistances in series, is dominated by those whose values are largest. Thus improvements in already large values of h will have very little overall effect if the other value of h is small and dominant.

In developing the concept of a heat transfer coefficient, we moved away from physics and into the realm of engineering science. Equation (4.15), which is applicable to a clean system, was developed. In it, allowance is made for the thermal resistances within the fluid films and the resistance of the intervening tube wall. However, all industrial systems contain impurities and some will be deposited on the heat transfer surfaces in the form of thin dirt films, which being of low thermal conductivity will generate additional thermal resistances. These films build up with time and the heat exchanger must be sized so that the performance is satisfactory throughout the complete period of operation.

The incorporation of inside and outside fouling resistances into the design equation is easy, but the resulting equation may mask the uncertainty surrounding their estimation. No reliable predictive methods are available, and past experience is the best guide. We have thus moved away from an area dominated by engineering science to one which, in a small way, typifies an engineering task in which there is a need to make good estimates on the basis of incomplete knowledge. The full design equation for estimating the overall coefficient is obtained from equation (4.15) with the addition of two **fouling resistances**, f_i and f_o. They can be considered to be the reciprocal of dirt film coefficients, which, if the thermal conductivity and thickness of the films were known, would equal k_{film}/t_{film}. Incorporating these into equation (4.15), and using diameters instead of radii, the final form of the relationship (for a shell and tube exchanger) between the overall and individual coefficients, is given by:

$$\frac{1}{U_{in}} = \frac{1}{h_{in}} + f_{in} + \frac{d_{in}\ln(d_o/d_{in})}{2k_w} + \frac{d_{in}}{d_o} \times \frac{1}{h_o} + f_o \qquad (4.17)$$

Typical fouling resistances (and for ease of comparison their reciprocals, the so-called dirt film coefficients) are given in table 4.3. By looking at this and the preceding table, it can readily be seen that in certain cases (for example, condensing vapour on shell side with cooling water flowing through the tubes) the fouling resistances can

Table 4.3 Typical fouling factors (dirt film coefficients)

Fluid	Fouling factor $(m^2\ K\ W^{-1})$	Coefficient $(W\ m^{-2}\ K^{-1})$
Boiler feed water	0.0001 –0.0002	5000–10000
Cooling water (from cooling towers)	0.00017–0.00033	3000– 6000
River water	0.00015–0.0004	2500– 6700
Steam (oil free)	0.00005–0.00025	4000–20000
Air and industrial gases	0.0001 –0.0004	2500–10000
Organic liquids and liquid oils	0.0002 –0.0005	2000– 5000
Heavy hydrocarbons	0.0005 –0.0002	500– 2000
Boiling organics	0.0002	5000
Condensing vapours	0.0001	10000

be the largest individual resistance. Representative values of the overall coefficients for shell and tube exchangers are given in table 4.4. It is of interest to note the wide variation within and between the various classes.

Table 4.4 Typical overall coefficients for shell and tube exchangers

Type of duty	U $(W\ m^{-2}\ K^{-1})$
Condensing steam — organic liquid or light oil	250–1500
Condensing steam — heavy hydrocarbon	60– 500
Condensing steam — gas	20– 200
Superheated steam — gas	10– 50
Condensing organic — water	750–2500
Water — organic liquid or light oil	250–1000
Water — heavy hydrocarbon	60– 300
Water — gases	20– 200
Water — water	800–1500

Temperature Driving Force

Having obtained a feel for the quantity U, it is time to examine the other main quantity governing heat exchanger performance: the temperature gradient or driving force. The question we need to answer is 'If the temperature of fluids A and B are changing from θ_A^{in} and θ_B^{in} to θ_A^{out} and θ_B^{out}, what is the average temperature difference?'

Typical temperature profiles for tubular heat transfer are shown in figure 4.8 for both co-current and counter-current configurations. The heating and cooling curves cannot cross (why is this?) but for counter-current heat transfer the exit temperature of the initially cool stream can be hotter than the outlet temperature of the initially hot stream. For either configuration the terminal temperature differences might be 30 K and 10 K, but is the average driving force equal to 20 K? The answer is 'no'.

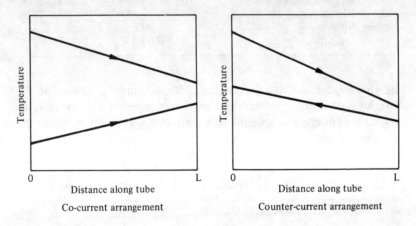

Figure 4.8 *Typical temperature profiles for tubular co-current and counter-current heat transfer*

The mathematical analysis is not covered in this introductory text but for both true co-current and counter-current flow the temperature driving force when properly averaged is found to be the logarithmic mean temperature difference between the fluids:

$$\Delta\theta_{lm} = \frac{\Delta\theta_2 - \Delta\theta_1}{\ln(\Delta\theta_2/\Delta\theta_1)} \qquad (4.18)$$

where $\Delta\theta_1$ is the temperature difference at one end and $\Delta\theta_2$ is the temperature difference at the other.

The particular temperature differences given earlier were $\theta_2 = 30$ K and $\theta_1 = 10$ K. The log mean temperature difference which results is 18.2 K. Thus the use of the arithmetic mean would have introduced a 10 per cent error. By inserting illustrative pairs of values such as (10, 15), (5, 25) and (2, 18) into equation (4.18) and comparing the resulting value of $\Delta\theta_{lm}$ with the respective arithmetic mean, it can readily be established that the arithmetic mean is always greater than the log mean.

Few shell and tube heat exchangers (including the one illustrated in figure 4.5) generate purely counter-current flow; cross-flow is generally present. Thus in design work the usual practice is to estimate the 'true' temperature difference by applying a correction factor F to the logarithmic mean temperature difference. The correction factor is dependent upon the temperature differences and the layout of the heat exchanger. Few engineers are concerned with the origins of the correction factors which are readily available in graphical form.

SIZING AND TYPES OF HEAT EXCHANGER

We are now in a position to frame an answer to the question posed at the beginning of the section on thermal conductivity. The general equation for heat transfer across a surface is $q = UAF\Delta\theta_{lm}$. For a given duty q, a certain type of exchanger would be chosen and values of U, F and $\Delta\theta_{lm}$ calculated. The area A would then be calculated and the physical layout specified. The layout affects the fluid velocities and so the values of U and F would have to be checked. If the calculated values differ significantly from the estimated values, the former are substituted for the latter and the calculation repeated. The procedure is thus seen to be iterative, which is the case with many engineering calculations.

The most widely used heat exchangers in the chemical and most other process industries are the various shell and tube units. The main type has already been shown in figure 4.5. The advantages of shell and tube heat exchangers are:

(a) Good mechanical layout; good shape for pressure operation.
(b) Can be constructed from a variety of materials, including glass and plastic.
(c) Uses well-established fabrication techniques.

Figure 4.9 *Schematic diagram of, and the flows within, a plate heat exchanger*

(d) Easily cleaned if straight tubes used.
(e) Reasonably large surface area per unit volume.

Generally, the materials of construction are metal. However, a few companies offer glass and 'Teflon' exchangers for use with particularly corrosive fluids. Another special type for use with corrosive fluids is the graphite block heat exchanger, which is literally graphite blocks with a matrix of holes. The fluids are generally in cross-flow and although there is a relatively large amount of 'wall' material between the fluid streams, the thermal conductivity of graphite is high and the resultant thermal resistance small.

Plate heat exchangers are of increasing importance. The simplified diagram (figure 4.9) shows that the hot and cold fluids flow counter-currently. The exchangers consist of a frame, a fixed end plate and a movable plate between which corrugated metal plates, gasketed at the edges, are fixed (figure 4.10). A port at each corner is either open or blind, according to need. A variety of gasket materials is used and the temperature range is now −40 to 200°C. Operating pressures of up to 25 bar can also be accommodated.

The modular design enables the exact size and number of plates to be used according to the level of heat transfer required. The heat exchangers are compact and for many duties a plate heat exchanger in stainless steel is cheaper than the equivalent shell and tube unit in carbon steel. While plate heat exchangers were originally developed and used as coolers in the dairy industry, they are now much more versatile. Large types can handle flow rates of 400 tonnes of liquid per hour, while at the other end of the scale, small units recover heat in laundries, heating and ventilation systems and swimming pools.

Figure 4.10 *A plate heat exchanger as assembled for use*

RADIATION

Radiation acts independently and in addition to conduction and convection. All surfaces emit energy in the form of electromagnetic waves and the governing equation for the net rate of exchange by radiation is

$$q_{rad} = A\epsilon\sigma(\theta_s^4 - \theta_{sur}^4) \qquad (4.19)$$

where σ = Stefan–Boltzmann constant
θ_s = temperature of surface (K)
θ_{sur} = temperature of surrounding (K)
and ϵ = emissivity.

It is important to use absolute temperature in the above equation.

The emissivity is a measure of a property of the surface and indicates how efficiently the surface emits radiation compared with an ideal radiator. Black bodies have a value of unity. Metallic surfaces generally have low values with oxide layers increasing their emissivity to around 0.3. Non-conductors have values of about 0.6.

Although equation (4.19) is highly dependent upon temperature, a common misconception is that radiation is only important at high temperatures. The heat loss from a pipe with a surface temperature θ_s which might typically be 50°C is given by the following equation that includes both conductive-convective and radiation terms:

$$q = h_0 A(\theta_s - \theta_{air}) + \epsilon A \sigma (\theta_s^4 - \theta_{air}^4) \qquad (4.20)$$

Commonly the outside heat transfer coefficient h_o and the term associated with it will be relatively low. Thus the radiation term which will be low can nevertheless, by comparison be significant. Exercise 4.4 illustrates this.

ENERGY RECOVERY

Although it is rare for the energy recovered to be as valuable as the product (though this is the case with sulphuric acid plants) the extraction of heat from hot streams is important because energy costs are often significant. Simple heat exchange will be covered first. Later an outline of a simple but powerful technique for multiple streams, a development of the late 1970s/early 1980s, will be introduced.

Consider two process streams S_1 and S_2. The one requiring energy, S_1, could be heated, for example, by steam, and the one whose temperature needs reducing, S_2, could be cooled by water. Both steam and cooling water need to be paid for and minimisation of their use reduces production costs. Alternatively it may be possible to exchange heat between the two streams S_1 and S_2. If S_1 needs to be heated from 50 to 100°C and S_2 needs to be cooled from 90 to 35°C, some of the heat that S_1 requires can be supplied by S_2 which reduces both the steam and cooling water usages.

In a simple single heat exchanger, two process streams, S_1 and S_2, can exchange energy provided heat can flow down a temperature gradient. Thus if stream S_1 needs to be heated to 100°C, the stream S_2 which is initially at 90°C can supply some of the heat. The final temperatures depend upon the relative flowrates, specific heats and the temperature approach, as shown in the following example.

Example 4.1

Consider a counter-current heat exchanger. The stream on the tube side has a specific heat capacity of 2.0 kJ kg^{-1} K^{-1} and has an inlet temperature of 100°C. The other stream enters at 35°C and has a specific heat capacity of 2.5 kJ kg^{-1} K^{-1}. The respective flowrates are 1.5 kg s^{-1} and 0.9 kg s^{-1}. Given an overall heat transfer coefficient of 800 W m^{-2} K^{-1} calculate the exit temperatures, and heat exchange area required, if 50 per cent, 75 per cent and 95 per cent of the maximum possible amount of heat is recovered.

Solution

Firstly it is important to understand what limits the recovery to a certain maximum. For every 10 kW of energy transferred the temperatures of the streams change as follows;

Stream S_1: reduction in temperature = 10/(1.5 × 2.0) = 3.33°C

Stream S_2: increase in temperature = 10/(0.9 × 2.5) = 4.44°C

Thus the cool stream S_2 approaches the temperature of the hot stream more quickly than the temperature of that stream approaches the temperature of stream S_2, as illustrated in figure 4.11. Remembering that a temperature gradient is required for heat transfer to occur, the limit in this example is stream S_2 reaching a temperature of 100°C. The maximum amount of heat recovery is thus:

Mass flowrate × Specific heat of limiting stream × (Maximum temperature change)

that is

0.9 × 2.5 × (100 − 35) = 146.25 kW

When the fraction of energy recovered is less than this, the exit temperature of stream S_2 can be calculated from:

Heat load (kW) = Mass flowrate × Specific heat × (Exit temperature − 35)

The exit temperature of the hot stream can also be calculated by way of a very similar equation. (What form does it take?) Then all of the temperatures will be known and the temperature driving force can be calculated. The overall heat transfer coefficient U has been

83

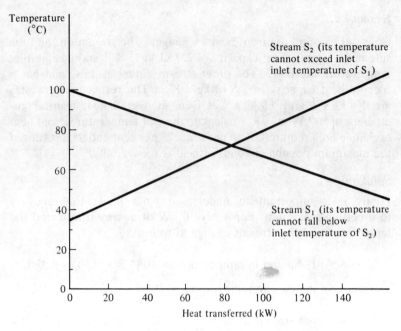

Figure 4.11 *Change in stream temperatures as a function of the amount of heat transferred. Comments refer to* counter-current *flow*

given and so the area of the exchanger can, as shown, be calculated.

Percentage recovery	50	75	95
Energy recovered (kW)	73.1	109.7	138.9
Exit temperature of S_1 (°C)	75.6	63.4	53.7
Exit temperature of S_2 (°C)	67.5	83.75	96.75
Temperature difference at hot end (°C)	32.5	16.25	3.25
Temperature difference at cold end (°C)	40.6	28.4	18.7
Temperature driving force (°C)	36.4	21.8	8.8
Area required (m^2)	2.5	6.3	19.7

The above example illustrates that (a) when *two* streams are matched, *one* of the streams limits the maximum theoretical recovery, and (b) the amount that can be recovered economically is often well below 100 per cent because the area required increases rapidly as the percentage recovery becomes high.

It follows from (a) that the selection of streams to pair together is a task that must be undertaken carefully. If the number of streams that need heating and cooling increases beyond a few, there are many ways of combining them into pairs.

A technique that is conceptually straight-forward, and attractive to engineers who are able to obtain a 'feel' for the problem follows. Those finding it a little too advanced can move straight to the next chapter because subsequent sections do not require an understanding of it.

Imagine that four streams have been identified as needing heating and that five streams need cooling. If the necessary enthalpy data are available, a curve representing the relationship between the temperature of the cold streams that are in need of heating and the respective enthalpy increases can be drawn. A simple example is shown in figure 4.12 which should be studied.

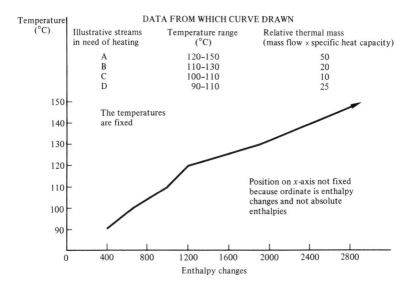

Figure 4.12 *Curve relating temperature to enthalpy changes for all streams that need heating — often called 'cold composite' curve*

A similar curve can be drawn for the enthalpy decreases and temperature levels of the hot streams that are cooled. When the two curves are placed on the same figure, their relative position with respect to each other is *not* fixed. While the temperature levels on the y-axis are fixed by process considerations, the enthalpy *changes* on the x-axis are not related to standard enthalpies but solely to enthalpy differences. Thus the process designer is free to select the most

85

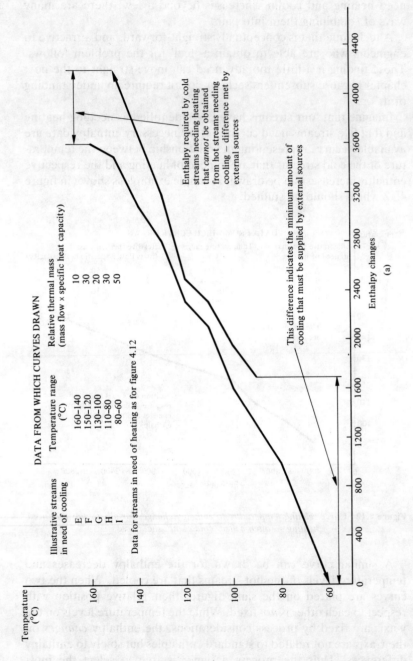

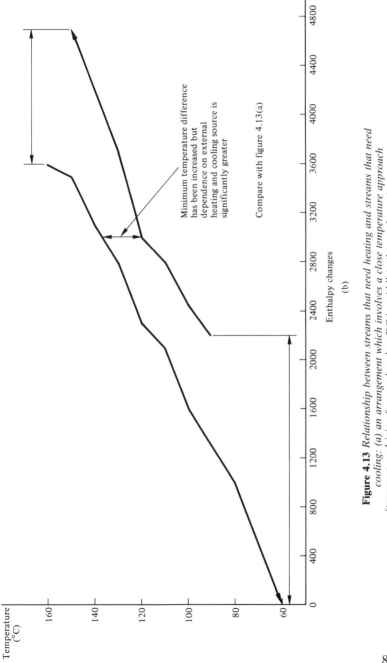

Figure 4.13 *Relationship between streams that need heating and streams that need cooling: (a) an arrangement which involves a close temperature approach (temperature driving force reduced to 3°C in middle); (b) an alternative arrangement*

appropriate horizontal displacement of one curve with respect to the other. Figure 4.13 illustrates the consequences of moving the curves apart: the overlap of the two curves decreases but the average temperature difference between them increases. The main consequence is that less heat can be transferred between the streams in need of energy and the ones needing to lose energy.

There is a limit to the proximity of the two curves because a temperature difference must exist if heat is to be transferred. It is common for design engineers to specify a minimum temperature difference of 10°C but on occasions it is economical to have temperature approaches of only 3°C, in which case very accurate prediction of enthalpy data is vital.

If the curves are moved until they touch, the point at which they touch is called the 'pinch' point. It has been established that heat exchanger network designs which have hot streams above the 'pinch' transferring heat to cold streams below the 'pinch' are wasteful of energy. The current approach is to identify the 'pinch' point and to split the energy integration problem into two problems: one an exercise in integration for streams above the 'pinch' (that is, establishing which streams should be paired with which) and another for the streams below the 'pinch'.

This elegant technique can be applied both to the design of new process plant and the upgrading of existing plants. The term 'upgrading' refers to developments which either enable the rate of production to increase or allow different feedstocks to be handled. The example is of the latter type and it first appeared in the technical literature in 1981. A company separating crude oil into its component parts (fuel oil, gas oil, kerosene, naphtha) wished to increase the capacity of the unit, to enable it to handle different feedstocks. The redesign involved the heat exchanger network before the distillation column. The initial design (figure 4.14a) by an engineering contractor proposed the installation of a new furnace which would pre-heat the crude oil prior to the main furnace. However the application of the then new techniques outlined above showed that that design involved substantial transfer of energy across the 'pinch'. The gas-oil heat exchanger was not being used efficiently because of poor positioning within the network. Redesign (figure 4.14b) showed that if it were repositioned, a new furnace and its associated circuit were unnecessary. The related saving on capital outlay and furnace fuel was substantial.

In a small way it is hoped that this illustrates the creative nature of design. The analysis showed that the original design was deficient but

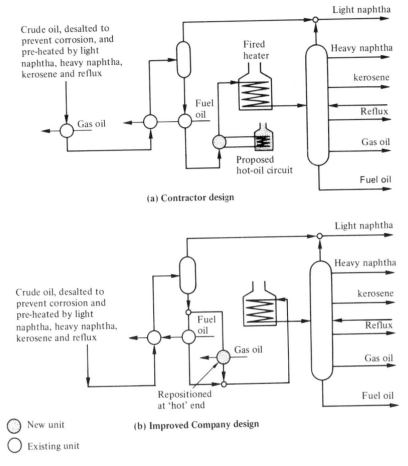

Figure 4.14 *Two possible heating exchange networks. Prior to distillation the crude oil is heated. The heat comes from the products that need cooling*

it did not point the way to one right answer; rather it suggested a target which the designer had to work towards.

EXERCISES

4.1. Estimate how much electricity or gas is likely to be saved over a period of one year if a living room is double glazed. Make explicit your assumptions and, besides quantifying in terms of thermal units, give monetary estimates.

4.2. Although heat exchanger design is a task in which only some process engineers will specialise, all process engineers will need to size heat exchangers. Using the information given in tables 4.2 and 4.3, estimate the surface area required for the following heat transfer operations. Also find the heat flux (heat load per unit area) in each case.

(A) The conversion of SO_2 to SO_3 is exothermic. The first stage of the catalytic reactor is designed such that the temperature of the gases at the outlet is 610°C. Before being passed to the second stage, the gases are cooled to 430°C; steam at a pressure of 20 bar (212.4°C) is produced. The gas flow is 8.6 kg s^{-1} and the average specific heat of the gaseous mixture is 1.1 kJ kg^{-1} K^{-1}.

(B) An organic product stream at 80°C is cooled to 30°C by counter-current heat exchange with cooling water available at 20°C. The flow rate of the product stream is 5 kg s^{-1} and the cooling water return temperature must not exceed 35°C. An average specific heat capacity of 2.1 kJ kg^{-1} K^{-1} is to be used for the organic stream which can be assumed to be non-fouling.

4.3. The comments in figure 4.11 refer to counter-current flow. If the two streams S_1 and S_2 entered the heat exchanger at the same end and flowed co-currently, what would the limiting changes in temperature be?

4.4. Estimate the rate of heat loss from a lagged 1000 m pipe with a surface temperature of 50°C. The pipe is 150 mm in diameter and is covered with lagging 100 mm thick and of emissivity 0.3. The following data are available:
Air temperature, typically 10°C
Stefan–Boltzmann constant = 5.67×10^{-8} W m^{-2} K^{-4}
Natural convection heat transfer coefficient for the outside of a pipe

$$= 1.18 \left(\frac{\Delta T}{d}\right)^{0.25} \text{ W m}^{-2} \text{ K}^{-1}$$

where ΔT = temperature difference (K)
d = outside diameter (m).

5 Separation Processes

It will be recalled that before and after most reaction stages, separation steps are necessary. Also certain processes have no reaction stages and are just a series of physical unit operations, for example, milk processing or natural gas treatment. The raw material may need purifying and with regard to the Case Studies (CS) the following are important examples: for CS1, filtration of molten sulphur and drying of the combustion air; for CS2, removal of sulphurous compounds from the feed to the primary reformers; for CS3, air purification is often important; and for CS4, water and silt, which are produced simultaneously with the oil and gas, are removed from the hydrocarbons as soon as possible.

Prior to storage and subsequent sale, separation of desired products from by-products (if any) and from waste streams is obviously necessary. The following examples have already been encountered: separation of gases by selective absorption of one gas into a liquid, such as removal of sulphur trioxide from the waste gases (CS1), and removal of carbon dioxide from the synthesis feed stream on the ammonia plant (CS2); separation of solids from liquids, for example, removal of bio-mass product from fermenters (CS3); and separation of liquid from gases such as that of condensed ammonia from uncondensed nitrogen and hydrogen (CS2). Other examples linked to the Case Studies will be encountered later, but firstly, the range of separation processes is systematically reviewed.

RANGE AND CHOICE

It is important that the process engineer has the ability to select and then size the appropriate separation process. The production rate of a plant is as high as the production rate of the weakest link within the overall process. Thus it is important that all pieces of equipment from pumps to reactors to pipe lines to separating vessels are sized correctly.

Tables 5.1 to 5.4 reveal the wide range of available processes for the purification of reactants or products. In order to achieve a separation it is necessary that the component parts of a process stream possess different properties that can be exploited by a particular operation. The thermal processes (for example, distillation, evaporation and crystallisation) depend on differences in vapour pressure or solubility. In some cases the separating agent is not heat but another fluid or solid (for example, a non-condensable gas for stripping out volatile liquid components from less volatile

Table 5.1 Separation processes for liquid mixtures

Name	Type of process	Practical examples/comment
Distillation	Thermal: depends upon the liquids having different vapour pressures	Separation of crude oil into component parts (see chapter 6)
Stripping	An insoluble gas is bubbled through a liquid and the volatile liquids are removed in the gas stream	Removal of light hydrocarbons from crude oil fractions
Solvent extraction	An immiscible solvent is contacted with the liquids that need separating. One or more, but not all, are soluble in the second solvent	Separation of aromatics from paraffins and naphthenes (cyclic non-aromatics)
Adsorption	The liquids are passed through a granular bed of solid. One of the liquids is retained upon the solid and removed later. Process depends upon differences in adsorption potential	Removal of trace amounts of water from hydrocarbon streams. Important if subsequent stage requires very low water levels. Separation of aqueous ethanol, close to the azeotropic composition, into pure water and pure ethanol
Gravity settling	For immiscible liquids the density difference may yield sufficient separation	Oil–water separation
Chromatography	Consists of a mobile phase and a stationary phase. Components of the mobile phase attach themselves to the stationary phase, but to different extents. As the retardation times differ, a separation of the components occurs	Principally used for analysis; but protein separations and other biotechnological separations need separators of this type within the manufacturing process itself

Table 5.2 Separation processes for liquid–solid mixtures

Name	Type of process	Practical examples/comment
Evaporation	Concentration of solutions by boiling off solvent. The vapours above the solution are pure solvent. Contrast with drying	Concentration of milk. Concentration of fruit juices. Concentration of sodium hydroxide solutions (chlorine–alkali plant). Evaporation of sea-water to produce potable water
Drying	Removal of a volatile solvent from a solid in the presence of a non-condensable gas, normally air	Wide range from drying of crystals to drying of biscuits
Crystallisation	Solubility limit of desired solid is exceeded either by evaporating some of the solvent or simply by cooling	Production of raw sugar from the clarified juices. Separation of *para*-xylene from *ortho*- and *meta*-xylene. Discussed in text
Precipitation	Addition of chemical reactant to give insoluble precipitate	Waste-water treatment
Sedimentation	Depends on density difference. Moderate sized particles will separate out in a reasonable time	Removal of silt from oil at well-head
Filtration	Depends on size of solid being greater than pore-size of filter	Widely used to protect equipment such as pumps and nozzles (for example, burner nozzles) and to recover solid product from liquid (for example, cells from fermentation liquor or waxes during processing of crude oil
Centrifugation	Imposition of larger gravitational forces aids separation of particles from liquid. Important if density difference small and/or particles small	Widely used, particularly in pharmaceutical industry. Crystals often concentrated in centrifuges, for example, *para*-xylene and sugar
Microfiltration	An extension of normal filtration to the micron level	Discussed in separate section
Ultrafiltration	An extension of normal filtration to the sub-micron level. Sieving on a molecular scale	The concept of pores within the filter is now inappropriate. Discussed in separate section
Dialysis	Selective membrane: separation occurs because of different diffusion rates which depend on molecular weights	Artificial kidneys. Recovery of sodium hydroxide in rayon manufacture
Electrostatic precipitation	Very fine droplets can be charged and then attracted to collecting plates	Has been used to remove sulphuric acid mist from gaseous effluent. See discussion

Table 5.3 Gas–liquid and gas–gas separation processes

Name	Type of process	Practical examples/comment
Gas–liquid		
Stripping	Removal of a dissolved gas from a solution by bubbling an inert gas through the liquid	Removal of carbon dioxide from liquid by bubbling air through the solution
Gravity separators	Separation of gas from liquid by allowing sufficient time for (a) droplets of liquid to fall out of gas, and (b) bubbles of gas to rise through liquid	Gas–oil separators (CS4). 'Knock-out' vessels in front of compressors; the fuel gas for gas turbines must be free of liquid droplets and these drop out in 'Knock-out' vessels which are described later
Cyclones	Depends on density differences. Liquid droplets thrown to wall	Removal of acid spray from gaseous effluent leaving the absorbers on a sulphuric acid plant (CS1)
Gas–gas		
Absorption	Preferential absorption of one component into a liquid	Removal of SO_3 from other gases (CS1). Removal of CO_2 from synthesis gas (CS2)
Adsorption	Preferential adsorption of one component on to a solid surface	Removal of water vapour by silica gel or molecular sieves. Odour removal
Membranes	Small molecules diffuse through the membrane	Separation of hydrogen from other gases (for example, purge gas (CS2))
Diffusion	Different diffusion rates through porous barrier	Concentration of radioactive $^{235}UF_6$ from natural UF_6
Thermal diffusion	Different diffusion rates along a temperature gradient	Isotope separation

liquid components, a solvent for extraction or a solid adsorbent). A further class of separations is based on either size differences, density differences or a combination of both.

On studying the tables, it will be seen that there is a number of links between them. For example, knowledge of how small bubbles rise through a liquid will be of use in the design of immiscible liquid–liquid and solid–liquid gravity settlers as well as gas–liquid gravity separators, since all three depend upon the same basic principle: namely, the application of Newton's laws of motion to a freely moving body.

For a small spherical particle of diameter d and density ρ_s settling

Table 5.4 Gas–solid and solid–solid separation processes

Name	Type of process	Practical examples/comment
Gas–solid		
Cyclones	Depends on density differences. Solid particles thrown to wall	Recovery of fine solids from gas streams
Electrostatic precipitation	Fine solids are charged and then attracted on to collecting plates	Dust removal from stack gases. Often used at cement works
Filtration	Size of solid greater than pore-size of filter	Removal of dust
Microfiltration	Removal of harmful bacteria	Sterilisation of air
Solid–solid		
Freeze drying	A thermal process in which ice is sublimated to water vapour to leave a water-free solid	Dehydration of food
Solvent extraction (leaching)	Preferential dissolution of one solid	Purification of penicillin. Recovery of sugar from crushed cane or beat
Supercritical gas extraction	Depends upon solubility of a component in a gas; variable solubility with pressure permits recovery of component and re-use of gas	De-caffeination of coffee using CO_2
Flotation	Surfactants and air bubbles required. One solid preferentially adheres to surfactant which is carried to surface by bubbles	Ore flotation, for example, recovery of ZnS
Magnetic separation	Attraction of materials in a magnetic field	Concentration of ferrous ores

in (or rising up through) a fluid of density ρ, the net weight or buoyancy force is

$$F = \frac{\pi d^3}{6} (\rho_s - \rho)$$

Any initial acceleration of the particle is quickly dampened and the counter-balancing drag force is given by

$$F = 3\pi \mu d U$$

where U is the terminal velocity and μ the viscosity.

Eliminating F from the above equations, the equation for the terminal velocity is found to be:

$$U = \frac{(\rho_s - \rho)gd^2}{18\mu}$$

This is, strictly speaking, only applicable to rigid spheres (but in fact most liquid droplets behave as if they are rigid). It is also subject to the restriction that the Reynolds number $U(\rho_s - \rho)d/\mu$ is less than 1.0. Provided the restriction is met, the same simple expression holds for gas–solid, liquid–solid, gas–liquid and immiscible liquid–liquid systems, and is independent of the chemical nature of the process fluids and solids. This example illustrates both the need for having an understanding of fluid mechanics and the powerful nature of the unit operations approach, which enables knowledge gained in one area to be applied to another.

For larger Reynolds numbers the flow around the particles is turbulent and the simultaneous solution of two equations, which again are independent of the chemical nature of the system, is required.

In order to highlight some separation processes further, consideration is given to all of the Case Studies. A definite process has not yet been chosen for Case Study 3, and so the application of membranes to the food industry will be the focus of the section on membranes and membrane processes. The concluding section is on process selection while the next chapter focuses in some detail upon the unit operations of distillation and absorption, including the sizing of distillation columns.

Case Study 1 Revisited

A chemical engineer concerned with sulphuric acid production would be aware of all the processes given in the tables but most familiar with only those of direct concern. These include filtration, drying and acid spray/mist removal.

The filtration of molten sulphur is necessary in order to remove the small amount, less than 0.1 per cent, of solid matter. Any solid matter left may lead to deposits in the combustion chamber and the **waste heat boiler** that follows it. However the resultant ash is mainly deposited on the first catalyst bed in the form of very fine solids. These gradually accumulate and the resultant rise in pressure increases power consumption and eventually the back-pressure limits the capacity of the main blower and thus the daily production rate. In order to minimise this potential reduction in throughput, the sulphur is filtered in a pressure leaf filter comprising a large number of stainless steel wire-gauze elements. When properly prepared, the solids level is reduced to a value of about 0.001 per cent.

Figure 5.1 schematically shows the build-up of a 'filter cake' on the upstream side of a filter. The cake is often less porous than the filter itself, in which case the cake severely restricts the flow rate. This is true of the unwanted solids in molten sulphur and in order to overcome this problem a **filter aid**, in this case a diatomaceous earth, is added to the feed stream. The filter aid is retained upon the filter together with the unwanted solids, but the two together give an open porous structure which only imposes a small restriction upon the flow.

In a typical operation 50 kg of filter aid and 25 kg of lime, for corrosion protection, are mixed with 5 tonnes of molten sulphur and this mixture is then circulated through the filters to 'pre-coat' them. Once this operation is complete, the filters can be used to filter 750

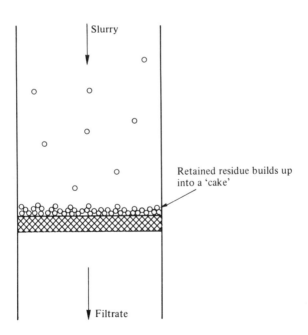

Figure 5.1 *Schematic diagram of filter in 'dead-end' mode*

tonnes of sulphur. With other types of filter such as rotating drum filters (in which a vacuum on the inside of the drum draws the filtrate through) it is possible to remove continuously the top layer of filter cake. This is important if the amount of solids to be removed is relatively large.

The filtered air to the combustion chamber is dried in order to minimise corrosion and the formation of acid mist. It is the presence of water vapour in the gas stream that causes the problem. It combines with sulphur trioxide in the reactor to produce sulphuric acid vapour which on cooling condenses to form a stable mist, an undesirable proportion of which would pass through to the atmosphere. In order to minimise the amount, the combustion air is passed through a drying tower. Concentrated sulphuric acid flows downwards across ceramic solids while air that has been moderately compressed in the air blower flows upwards. Most of the water vapour becomes absorbed into the acid.

The air blower produces a pressure of about 0.3 bar above atmospheric pressure and this is sufficient to push the air and all of the product gases formed through the drying tower, the combustion chamber, the reactor, the absorbers, the filters in the top of the absorbers and up the chimney stack at the end of the plant. Quite clearly, techniques to calculate pressure drops are important because the correct sizing of the inlet air blower is vital.

In addition to the mist already mentioned (size range 0.01–3 μm) the exhaust gases, which are mostly the unused nitrogen and the excess oxygen, carry forward the spray formed in the absorbers. The spray is formed of droplets in the size range 5–100 μm and these are easily removed. The method is not a conventional filtering process but one of impingement. The basic principle is outlined in figure 5.2. The inertia of the particles approaching the rod is such that not all of them are carried around the rod by the gas. A single layer of rods would not capture sufficient droplets and a modern steel mesh demister consists of knitted stainless steel wire, 0.25 mm in diameter, crimped and built up into a pad 150 mm deep. It is like a giant 'Brillo' pad. The droplets that impinge upon the wires coalesce together and form sizes sufficiently large to fall back into the process liquid, as shown in figure 5.3. The vapours passing through the pad still contain the mist particles.

The mist can be removed by electrostatic precipitators which firstly impose a negative charge upon any particles and secondly attract them to an earthed plate. The mist particles on the plate coalesce and drain into a suitable tank below. Although electrostatic precipitators

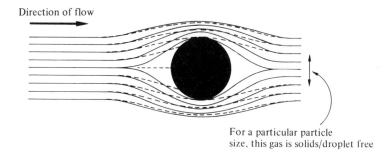

Figure 5.2 *Removal of particles (solid or liquid) by impingement*

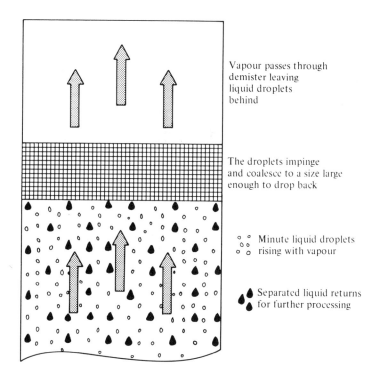

Figure 5.3 *Operating principles of a spray removal unit*

have a high efficiency and a low pressure drop, they are extremely expensive to install. Corrosion by acid also leads to high maintenance costs.

The alternative preferred today is the candle filter shown in figure 5.4. The physical process is one of impingement but because the glass fibre is only one-hundredth of the diameter of the crimped stainless steel wire, the compressed glass fibre candle filters are able to remove the fine mist. They have proved to be relatively cheap, efficient and reliable. Their main disadvantage is pressure drop which can be up to 0.1 bar. For a 2000 tonne per day plant, the extra power consumption is around one megawatt.

Case Study 2 Revisited

The hydrocarbon feedstock to an ammonia plant needs to be desulphurised. For a liquid feed such as naphtha the overall reactions can be represented by the following equations:

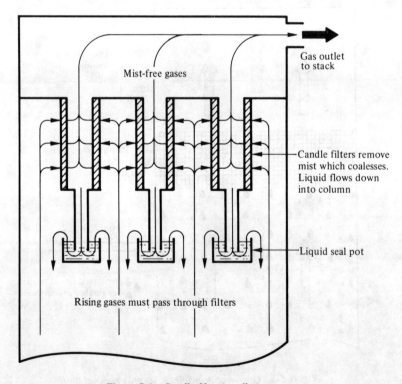

Figure 5.4 *Candle filter installation*

$$R\text{—}SH + H_2 \xrightarrow{\text{Co/Mo catalyst}} R\text{—}H + H_2S$$

$$H_2S + ZnO \longrightarrow ZnS + H_2O$$

Only the second equation is applicable for natural gas feeds. The amount of hydrogen sulphide produced per day is relatively very low and a moderately sized vessel contains sufficient zinc oxide to adsorb all of the sulphurous gas that will be generated in one year. The spent adsorbent is simply removed and the vessel filled with fresh zinc oxide. This is rare. Normally a solid adsorbent (or liquid absorbent) is regenerated and re-used. This is certainly necessary in the absorption process that is used to remove the carbon dioxide; over 1 tonne of carbon dioxide is produced for every tonne of ammonia.

The solubility of carbon dioxide in water is low even at 25 bar, and so the acid gas may be removed by absorption and reaction with a solution of potassium carbonate and bicarbonate. The overall reaction is:

$$K_2CO_3 + CO_2 + H_2O \rightleftharpoons 2KHCO_3$$

Alternatively an amine solution, generally monoethanolamine ($NH_2CH_2CH_2OH$), can be used as the absorbent. The principal reactions occurring are as follows:

$$2RNH_2 + CO_2 + H_2O \rightleftharpoons (RNH_3)_2CO_3$$
$$(RNH_3)_2CO_3 + CO_2 + H_2O \rightleftharpoons 2RNH_3HCO_3$$
$$2RNH_2 + CO_2 \rightleftharpoons RNHCOONH_3R$$

The primary and most rapid reaction is the latter.

Absorption takes place in large columns containing either trays or packing as described in the next chapter. The carbon dioxide-rich liquid stream leaving the bottom of the absorber (see figure 5.5) is piped to a regenerating column operating at a pressure just above atmospheric. The reduction in pressure, together with heating and the stripping action of the steam, removes the absorbed carbon dioxide from the liquid. The regenerated absorbent is then cooled and recycled by pump to the absorption column.

The above stages are some of the preparatory stages for the production of the nitrogen plus hydrogen feed to the ammonia converter. In the synthesis loop (figure 1.5) the ammonia is separated from the unconverted gases by condensation and gas–liquid separa-

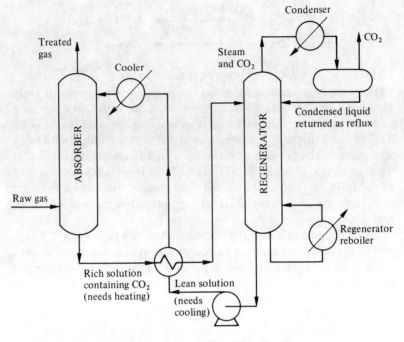

Note: (a) Heat exchange between rich solution stream and lean solution stream.
(b) Reflux on regenerator (reflux is discussed in chapter 6).

Figure 5.5 *Carbon dioxide removal system, with continual regeneration of absorbent solution*

tion. Given the differences in boiling point, this is a convenient and obvious process to choose. As explained in chapter 2, a purge stream is essential in order to avoid continual accumulation of argon in the reactor. The synthesis loop feed also contains methane, and as this also is an inert, with respect to the synthesis reaction, it too would accumulate. Until recently, there was no effective and economic means of separating the unwanted gases in the recycle stream from the reactants in that stream.

The relative molecular masses of the gases in the purge stream are 2 (hydrogen), 28 (nitrogen), 40 (argon) and 16 (methane). The hydrogen will diffuse through membranes which effectively exclude the other gases. An economically feasible process has only been available since 1979 because the membrane must be very thin if the flow of hydrogen is to be sufficiently large. The new gas separation membranes consist of a very thin leak-free homogeneous layer of a highly gas-permeable polymer deposited upon a fairly porous supporting substructure. The employment of this process enables the hydrogen to be recovered and recycled. The alternative, which has been practised for many years, is to pipe all of the purge stream, untreated, to the boiler house and burn it. In this way the fuel value of the stream is realised and it is disposed of safely.

Case Study 3 Revisited

In the **fermentation** process for the production of bio-mass, which was introduced earlier, the key separation is the recovery of the cells from the nutrient liquor. Clearly, a filtration process is involved. The fermenter itself is operated under sterile conditions and it would be desirable to maintain sterility at the filter because, as shown in figure 5.6, this would enable the returning liquor to be within the sterile barrier. If this were to prove impossible, the returning stream would need to be sterilised because it is important to exclude foreign bacteria from the fermenter. Thus process designers in this area have an extra challenge rarely met in traditional industry or on university courses.

Case Study 4 Revisited

Off-shore, the amount of processing is kept to a minimum and consists mainly of separating the well-head stream into water, oil and

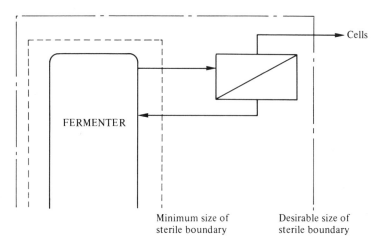

Figure 5.6 *Sterile boundaries: a challenge unique to biochemical processes*

gas. Water is always present and the percentage in the feed increases steadily with time. The ratio of oil to gas is highly variable; on some oil fields there is barely sufficient gas to provide fuel for the platform (the platform aims to be self-sufficient), while with other fields the gas is as important a product as the oil. In order to conserve the limited fossil fuel resource, most governments now prohibit excessive **flaring** of gas. If it is not recovered and piped ashore, it is re-injected into the reservoir.

The number of separating vessels will depend upon the reservoir pressure and the oil-to-gas ratio, but the first one is always a three-phase separator yielding water, oil and gas. The water will still contain some oil droplets and further treatment will be required before it can be discharged to the sea. The amount of treatment will depend upon the quantity involved and governmental regulations. The gas passes through a demisting pad, similar to that described above for acid spray removal, and is then added to the gas streams from the other separators for further treatment.

If the gas is to be compressed and transported by sub-sea pipe line, then it may be necessary to remove carbon dioxide. If only a few per cent is present it is cheaper to transport the unwanted gas and remove it on-shore than to remove it off-shore. If, as with one North Sea field, the amount is 20 per cent, then removal off-shore is undertaken. The absorption process uses monoethanoiamine and is basically the same as that on an ammonia plant. Absorption can also be used to remove water, and the re-usable absorbent which is cycled between absorption tower and regenerator is triethylene glycol. Water removal limits corrosion damage and prevents the formation of solid hydrocarbon hydrates which can plug gas lines.

MEMBRANES AND MEMBRANE PROCESSES

A membrane is a selective barrier between two homogeneous phases (gas or liquid). The selectivity is based on the fact that the different components in the feed will diffuse through the barrier at different rates. This is illustrated in figure 5.7 in which some of the terms are defined. It is worth remembering that, in general, a membrane is not a complete semi-permeable barrier and some of the solute, which should ideally be retained in the concentrate, passes into the permeate.

The first synthetic membranes were prepared at the beginning of the century and developed for laboratory scale use after the First

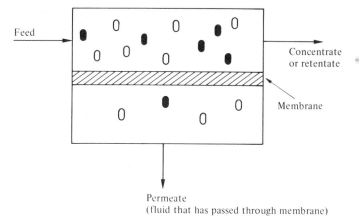

Note the 'cross-flow' arrangement. Compare with figure 5.1.

Figure 5.7 *Schematic representation of separation by a membrane*

World War. The membranes were composed of rather thick layers of cellulose nitrate or cellulose nitrate/cellulose acetate. Although technically feasible, separations were not economically feasible on the industrial scale at this time. The necessary equipment would have been excessively large because the amount flowing through the membrane per unit of membrane area per unit time (the 'flux') was very low. The major breakthrough came in the 1960s with the development of asymmetric membranes, schematically illustrated in figure 5.8. Figure 5.9 divides membranes into three main groups and

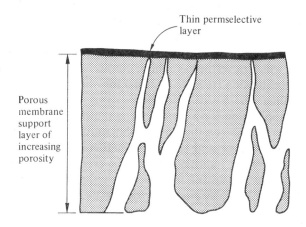

Figure 5.8 *Illustration of an asymmetric membrane*

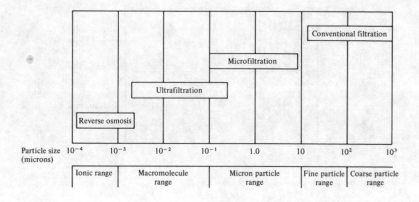

Figure 5.9 *Size range of various classes of membrane*

indicates the nominal pore-diameters of the various classes of membrane.

A thin barrier layer is essential for gas separations and should be dense, non-porous and selective. The passage of components through this layer is often modelled on the basis that the components dissolve in the top surface, diffuse across the membrane and desorb on the downstream side. On the other hand, the free space within microporous membranes is sufficiently large for fluids to flow through in a conventional manner. These membranes are generally still asymmetric in order to have only a small area of high resistance but lack the extra 'skin' shown in figure 5.8. Their pore diameters are in the range 0.1–10 μm, which means that they can be pictured by simple electron microscopes. The resolution of scanning electron microscopes is, however, insufficient to elucidate the structure of ultrafiltration membranes, which nevertheless exhibit porous behaviour. Figure 5.9 includes hyperfiltration (reverse osmosis) for which the main use has been in desalination.

A typical application for a microfilter is shown in figure 5.10. An advantage of these filters is that they enable sterility to be maintained at the separation stage and so contamination problems are minimised. In the beverage industry, they can be used for the recovery of yeast from brewery waste and so reduce the effluent problem, and in the sugar industry to clarify glucose syrup and beet sugar juice. These filters can be ceramic which conveys obvious advantages. Figure 5.11 includes both an end-on view of a tubular unit containing nineteen ceramic elements and a scanning electron microscope photograph of

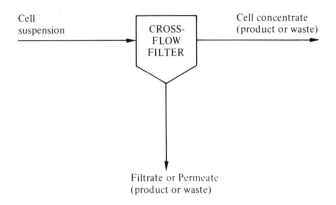

Figure 5.10 *Cell separation using a cross-flow microporous membrane filter. The main product can be either the cell concentrate or the filtrate*

a cross-section through the surface of a channel in one of the elements. The surface membrane is asymmetric.

The application of ultrafiltration membranes, which are currently all of a polymeric nature, is more widespread because it is possible to separate smaller molecules such as sugars from larger molecules such as proteins. The main attraction for ultrafiltering cheese milk is the increased yield that results from the incorporation of whey protein into the cheese. In a traditional process these proteins remain in the waste liquid whey. The waste stream from a membrane unit still contains lactose (milk sugar) and this can be used for alcohol production, as an animal feed or as a feed to an **anaerobic** digester which will produce methane. The technology is being applied to both hard and soft cheese; examples includes Cheddar and Camembert.

For yoghurt manufacture, the starter solution must be sufficiently concentrated and skim milk powder is often added back into skim milk prior to incubation. A cheaper alternative is to concentrate skim milk by ultrafiltration which has the added advantage that the lactose passes into the permeate.

Apart from the dairy industry, the main food and beverage areas of application are in fruit juice clarification and effluent pre-concentration. Unlike thermal processes, ultrafiltration allows ambient temperature separation and concentration, avoiding both high energy costs and degradation of heat-sensitive compounds. The units are compact and process control is straight-forward. For these reasons the range of applications given in table 5.5 can be expected to increase steadily.

Principle of the Ceramic Membrane Element

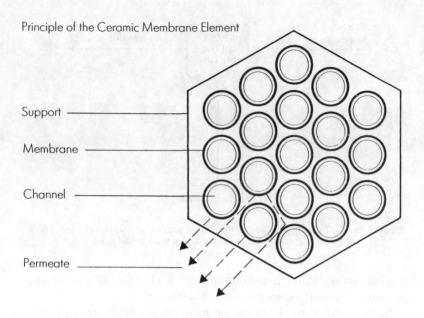

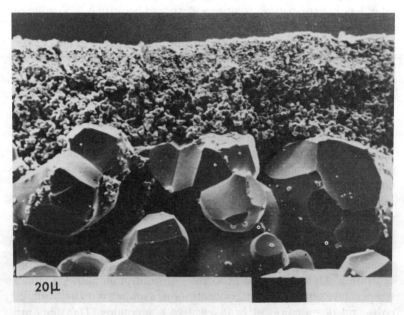

Figure 5.11 *Photograph of ceramic unit and scanning electron microscope photograph of membrane surface*

Table 5.5 Membrane processes: typical applications

Reverse osmosis (hyperfiltration)
Sea-water desalination
Recovery and recycling of indigo dye from waste-water
Concentration of pharmaceutical-grade sugar to 30 per cent
Concentration of amino acids, vitamins and natural extracts/flavours
Concentration of antibiotics

Ultrafiltration
Removal of pectin hazes and suspensions from fruit juices
Recovery of lignosulphonate and vanillin from pulp/paper effluent
Recovery of proteins, for example, from whey
Concentration of milk solids prior to cheese manufacture
Recovery of protein and enzymes from food processing effluents
Fractionation of blood plasma
Removal of pyrogens from water
Hydrogen recovery (special membrane required)
Electropaint recovery
Recovery of oil from oil–water emulsions
Waste-water treatment

Microfiltration
Concentration and separation of products from fermentation broths
Filtration of solvents
Clarification and cold sterilisation of juices, wine and beer
Recovery of beer from yeast sediment

PROCESS SELECTION

There is generally a number of technically feasible processes for any given separation. For example, a solution of common salt can be concentrated by freezing, by evaporation or by reverse osmosis. The skill of a process design team is to select an economically feasible process. The separation of xylenes is a classic example.

Para-xylene (1,4-dimethylbenzene) is an important petrochemical which is an intermediate for the manufacture of chemicals used in the production of polyester fibres, for example, terephthalic acid (benzene-1,4-dicarboxylic acid), $HOOC-C_6H_4-COOH$. It is one of the three xylene isomers. The term 'mixed xylenes' refers to these isomers and ethyl benzene:

ortho-xylene   *meta*-xylene   *para*-xylene   ethy benzene

They have similar vapour pressures and their respective normal boiling points differ by less than 8.5°C. A consequence is that the production of individual components of high purity by distillation is difficult, and with regard to one separation (para-meta), almost impossible.

	Ethyl benzene	*p*-xylene	*m*-xylene	*o*-xylene
Boiling point (°C) at 1 atm	136.19	138.35	139.10	144.41
Freezing point (°C)	−94.98	13.26	−47.87	−25.18

The process that has evolved includes crystallisation; *para*-xylene has a significantly higher freezing point. Figure 5.12 gives the overall flow diagram. The *ortho*-xylene is removed in column 1, together with the C_{9+} aromatics from which it is subsequently separated in another distillation column. The 5.3°C difference in boiling points between *ortho*-xylene and *meta*-xylene is not large and over 100 stages (a stage is described in the next chapter) are required. After removal of ethyl benzene, a feedstock for styrene production, the *meta*-xylene and *para*-xylene are cooled and *para*-xylene starts to crystallise out. Its recovery is limited because once the **eutectic** is reached, mixed crystals will form. When the liquor reaches a concentration of 87 per cent *meta*-xylene 13 per cent *para*-xylene, the practical limit has been reached and over 35 per cent of the *para*-xylene in the original feed stream is still uncrystallised.

Filtration of the mixture yields wet crystals which are further purified by removal of the mother liquor in a centrifuge. A final purity of 99.5 per cent is achieved. The filtrate which contains the unrecovered *para*-xylene is by weight 87 per cent *meta*-xylene, for which there is very little demand. This stream is isomerised in a reaction stage which results in the production of a mixture of all four

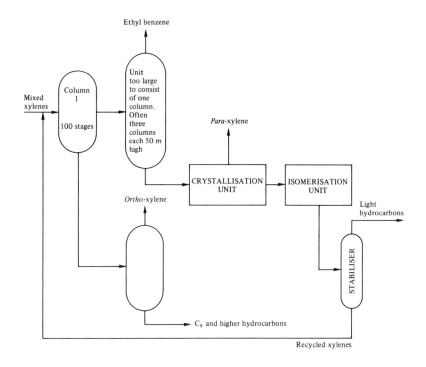

Figure 5.12 *Flow diagram of a xylene plant*

C_8 compounds. These are recycled, mixed with fresh feed and reprocessed.

The above example, which completes the overview on the range and choice of separation processes, involves a wide range of unit operations and indicates how different classes of separation processes can be combined to produce an economic plant.

However, technology has developed in competing areas and newer methods of separation such as adsorption now compete with the traditional energy-intensive crystallisation process outlined above. It is anticipated that most future plants will avoid the inclusion of a crystallation stage for the *para*-xylene/*meta*-xylene separation.

EXERCISES

5.1. If all of the catalysts used in the ammonia process could be replaced by catalysts whose performance was unaffected by the presence of hydrogen sulphide, would it still need to be separated out? If 'yes', suggest the type and location of the chosen separation stage.

5.2. In chemical engineering it is often necessary to make estimates using incomplete data. Estimate the relative effective capacity of the following two liquids for absorbing carbon dioxide: (a) monoethanolamine solution (20 per cent by weight), and (b) water at various pressures. The MEA absorptive capacity is in practice limited to 0.35 moles of carbon dioxide per mole of amine. The solubility of carbon dioxide in water at 20°C is about 0.039 mol litre^{-1} bar.

5.3. Find out more about the production of a common breakfast food or drink product such as margarine, sugar, instant coffee or cornflakes, and list the separation processes important to its production.

5.4. An aqueous solution of salts, macromolecules and microbes is fed across a reverse osmosis membrane, an ultrafiltration membrane and a microfiltration in that order. What do the three permeates contain?

6 Distillation and Absorption

The main emphasis will be upon stagewise, continuous feed distillation, schematically shown in figure 6.1. The column may contain trays or packing (as described later) to promote good vapour–liquid contact. The quantitative analysis is confined to two-component (binary) systems in trayed columns.

The vapour above, and in equilibrium with, most two component liquids is of a different composition from the liquid; the exception is **azeotropic** mixtures (see figure 6.2). Thus if, with normal mixtures, the vapour is separated off and condensed, a partial separation has been achieved; the condensed vapour is richer in the more volatile component than the liquid from which it came. The unit operation of distillation utilises this phenomenon, but it is important to remember that, unlike evaporation, the vapour normally contains a significant quantity of the less volatile component as well. The rising vapours are further enriched by contact with descending liquid. The resultant vapour is somewhat richer in the more volatile component than the original vapour, while the resultant liquid is somewhat richer in the less volatile component than the original liquid. Repetition of this process of contacting and physical separation, which is illustrated in figure 6.3 for trayed columns, gradually produces a chemical separation. The descending liquid, known as **reflux**, is absolutely crucial to the operation. Without it, very little separation is achieved, and an analytical proof of this fact is presented shortly.

The following relates directly to trayed columns. The number of stages in both the section above the feed and in the section between the feed and the bottom of the column is commonly in the range 5–25. The latter section is often called the *stripping* section because the more volatile material is being removed from, or stripped out of, the descending liquid. The former section is known as the enriching or *rectifying* section, because the purity of the ascending vapour is gradually enhanced. The overhead vapour passes through a condenser which may be designed to condense all or part of the vapour

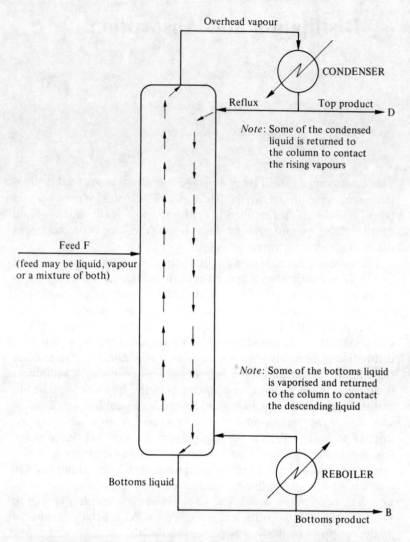

Figure 6.1 *Representation of a basic continuous distillation unit*

stream. The stream returned to the column, the reflux, must be a liquid, but the product stream can be either a vapour or a liquid. It is also possible to remove side-streams of intermediate composition and the classic example of crude oil distillation is illustrated in figure 6.4. A modern vacuum distillation column is shown in figure 6.5.

The need for reflux is illustrated by the following analysis. For a binary mixture, the mole fraction of the more volatile component in

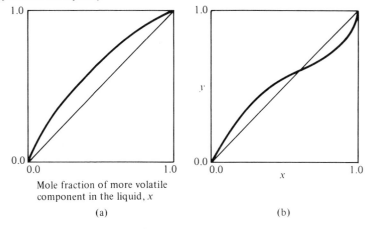

Figure 6.2 *(a) Typical y–x diagram for binary mixtures **not** showing azeotropy. (b) Typical y–x diagram for a minimum boiling point azeotrope*

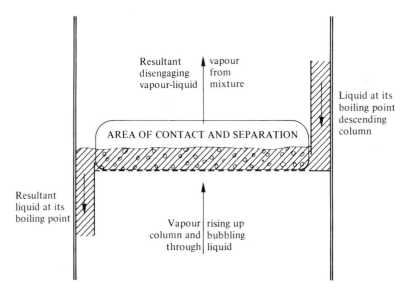

Figure 6.3 *Schematic diagram of a stage within a distillation column. Note that both mixing and separation are important*

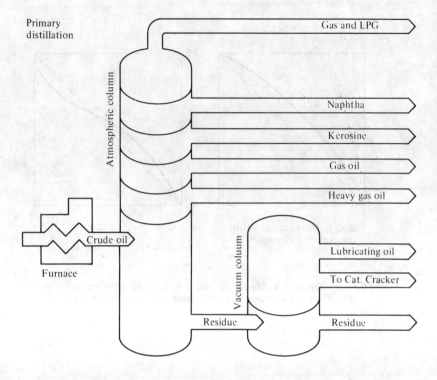

Figure 6.4 *Crude oil primary distillation: a wide range of products is produced from one unit. Most need further treatment before sale*

the liquid and vapour can be represented by x and y respectively. If the flow rate of vapour from the top plate (shown in figure 6.6) is V_T moles per hour, that from the plate below V_{T-1} and the corresponding liquid flow rates are L_R (reflux), L_T, etc., then the application of the steady-state material balance equations developed in chapter 2 leads to:

overall balance on top plate:

$$L_R + V_{T-1} = V_T + L_T$$

balance on more volatile component:

$$L_R x_R + V_{T-1} y_{T-1} = V_T y_T + L_T x_T$$

Figure 6.5 *Modern refinery vacuum distillation unit. One of the product streams is the feed to the plant producing lubricating oils*

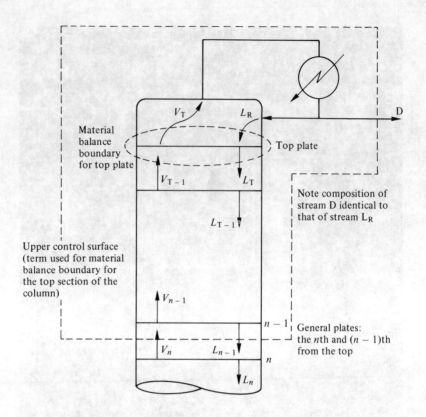

Figure 6.6 *Application of material balances to binary distillation analysis*

Assuming $V_T = V_{T-1} = V$ and $L_R = L_T = L$, which will be justified later, the balance equation can be written as:

$$y_T = y_{T-1} + \frac{L}{V}(x_R - x_T)$$

For a total condenser, the composition of the overhead vapour is the same as that of the reflex, that is, $y_T = x_R$, and so:

$$y_T = y_{T-1} + \frac{L}{V}(y_T - x_T) \qquad (6.1)$$

Given that y_T is greater than x_T (that is, the vapour arising from a liquid is richer than the liquid itself), the above equation proves that the vapour leaving a stage is (a) richer in the more volatile component than the vapour entering that stage, and (b) the extent of the change depends upon the ratio L/V, that is, the amount refluxed. With no reflux, $L = 0$, the composition of the vapour leaving the stage is the same as that entering the stage. Thus reflux is essential to the separation process.

If the vapour y_T is in equilibrium with the liquid x_T, the stage would be known as an *equilibrium*, *ideal* or *theoretical* stage. An actual stage in a distillation column will achieve between 40 and 90 per cent of the separation predicted for an equilibrium stage, because in part the contact time might be insufficient to obtain equilibrium. Also there will be incomplete separation of liquid and vapour, and some spray will be carried upwards in the vapour stream. The result of these inefficiencies is that if the design calculation shows that 30 equilibrium stages are required and the anticipated efficiency is 75 per cent, then at least 40 actual stages will have to be provided. A 10 per cent safety margin may be added and a unit with 44 stages built.

In the following sections, simple methods for choosing column pressure, for calculating the minimum reflux and the minimum number of stages, and for sizing columns will be introduced.

FACTORS AFFECTING DISTILLATION COLUMN DESIGN

Choosing the right pressure at which to operate a distillation column is very important. It has been established that reflux is essential, and so some of the vapours must be condensed and returned as reflux. Thus, it is generally desirable to have a column pressure such that it is a straightforward task to condense the overhead vapours.

On a refinery, a 'debutanizer' is a distillation column which removes the butanes from the higher hydrocarbons. As propane and lighter components have already been removed the distillate from a debutanizer is mainly normal-butane and 2-methylpropane(isobutane). Their normal boiling points are −0.5°C and −11.6°C respectively. So in order to condense butane vapours, the coolant in the condenser would have to be a refrigerant if the operating pressure were just atmospheric pressure. By operating at a pressure of around 5 bar, the condensation temperature is raised and water from a cooling tower or a river (suitably filtered and treated) can be used to perform the duty. The cost of using refrigerants far outweighs any increased capital costs arising from the installation of a higher pressure column and debutanizers are always operated at an elevated pressure with either water-cooled or air-cooled condensers.

'Depropanizers', which remove propane, are also part of the distillation sequence illustrated in figure 6.7, and they too are operated at an elevated pressure. With a pressure of about 15 bar, the temperature at the top of the column is sufficiently high for ordinary water or air to be used as the coolant for the condenser.

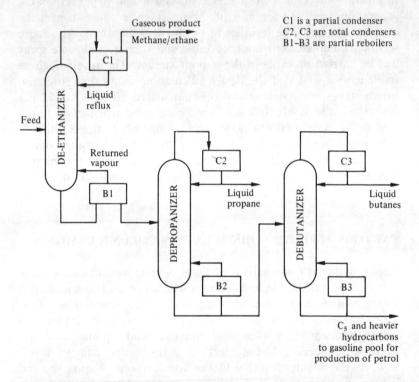

Figure 6.7 *Distillation sequence for production of propane, butanes and associated products*

The first distillation column in the sequence is a 'de-ethanizer' which removes methane and ethane. Although the latter can be cracked to ethene, the two are generally not separated and are used as fuel in a power station. This column too would ideally operate at a pressure such that the reflux was at about 30°C. However, as the pressure is increased an upper limit is reached. It is *not* a mechanical limit but one of a physical nature. Above its critical pressure a pure substance will exhibit just one fluid state irrespective of temperature. Mixtures also show similar behaviour but the critical point is labelled 'pseudo-critical' because it is not well defined. Now distillation is impossible without the counter-current flow of liquid and vapour and so a distillation column must operate at a pressure below the pseudo-critical pressures of the mixtures within the column. There is not just one pseudo-critical pressure, P_{crit}, but one for every mixture within the column. for a 'de-ethanizer', the top product has a P_{crit} value of around 70 bar, but the bottom product has a significantly lower P_{crit} value of less than 50 bar. It is necessary to work well away from this limit in order to ensure reasonable differences in densities between the liquid and vapour, and to have good control. An operating value of about 30 bar is chosen and this gives a temperature at the bottom of the column of about 105°C, which is, incidentally, 25°C below the temperature corresponding to the P_{crit} point. At this operating pressure, the temperature at the top of the column is about 0°C and so a refrigerant is required as coolant in the condenser.

Other factors affecting the design and performance of a distillation column are (a) the feed composition, (b) the specifications of the top and bottom products, (c) the number of stages in the column, and (d) the reflux ratio R. This ratio is the amount of reflux to the amount of distillate (top product). Both quantities are generally expressed in units of moles per unit time and therefore the reflux ratio is dimensionless.

The specifications of the top and bottom products crucially influence the required size of distillation column. The purer products require many more stages than run-of-the-mill grades. While five stages in the top part of the column might enable a 50 per cent mixture of A and B to be rectified into a mixture that is 90 per cent A, a further five might be required for an improvement of 6 per cent to 96 per cent, a further 5 for 3 per cent improvement to 99 per cent, and a further five for an improvement to 99.8 per cent. It is thus important to ascertain the exact requirements.

The feed composition, its temperature and whether it is liquid or vapour, are all important. A column designed for a liquid feed is unlikely to perform satisfactorily if the feed were to be all vapour. For given feed conditions and product quality there is a trade-off between the reflux ratio R and the number of stages N. However, there is both a minimum reflux ratio and a minimum number of stages and the next two sections will explain, with respect to binary mixtures, the origin of these minima.

Minimum Reflux Ratio

The concept will be introduced in the context of the McCabe–Thiele construction for the graphical analysis of binary distillation problems. This method uses y–x diagrams of the form shown in figure 6.2. Such diagrams are a graphical representation of equilibrium data, and so if the composition of the resultant vapour from an equilibrium stage is known, then the composition of the resultant liquid (in equilibrium with that vapour) can be obtained from the curve on the appropriate y–x diagram.

The curve represents the relative **volatility** of the two component across the composition range. At a particular composition the relative volatility α is given by

$$\alpha = \frac{y}{x} \bigg/ \frac{1-y}{1-x}$$

which on rearrangement gives an explicit equation relating y and x:

$$y = \frac{\alpha x}{1 + (\alpha - 1)x}$$

For close-boiling mixtures, α is effectively constant. More generally, α is a function (often linear) of x. Such data are to be found in reference books and computer databases. It is vital for computer-based solutions.

In order to calculate the required number of equilibrium stages, it is also necessary to have a second set of information linking the composition of the vapour leaving one stage to the composition of the liquid that is leaving the stage above. In figure 6.3 there are two pairs of such streams: 'resultant vapour disengaging from vapour–liquid mixture'/'liquid at its boiling point descending column' is one pair, and 'vapour rising up column'/'resultant liquid at its boiling point' is the second pair. The required relationship is given by the operating line equations. There is one for each section of the column, and the top operating line will be derived first.

Material balances taken over a material balance boundary or upper control surface shown in figure 6.6 on the overall amounts and on the more volatile components lead to

overall:

$$V_n = D + L_{n-1} \tag{6.2}$$

and on more volatile component:

$$V_n y_n = D x_R + L_{n-1} x_{n-1} \quad (6.3)$$

Elimination of V_n from the second equation and rearrangement gives:

$$y_n = \frac{L_{n-1} x_{n-1}}{D + L_{n-1}} + \frac{D x_R}{D + L_{n-1}} \quad (6.4)$$

This equation can be greatly simplified if the liquid flows from tray to tray are constant. 'Constant in what?' should be the question springing to mind. If measured in molal flows, it is reasonable, in many cases, to assume that the liquid flows (and hence the vapour flows) are constant within any section of the column. This arises from the fact that the molal latent heat of vaporisation, which is measured in kJ per mole, is often the same for both the more volatile and the less volatile components. Thus when this is so, it is true to say that for every one mole of vapour condensing on a given plate, one mole of vapour, albeit of a different composition, will be released.

For those cases where this simplifying assumption of 'equimolar overflow' can be made (and one of the first steps in any calculation should be to check the validity of the assumptions), equation (6.4) can be simplified. Also the vapour flow at any stage can be related to the liquid flow and the flow rate of the top product (all flows being in moles/unit time). Thus

$$V = D + L \quad (6.5)$$

Defining the reflux ratio R as the ratio of the amount of liquid returned to the amount of top product withdrawn, that is $L = RD$, equation (6.4) can be written as:

$$y_n = \frac{R x_{n-1}}{R + 1} + \frac{x_R}{R + 1} \quad (6.6)$$

This is the equation of the top operating line in which the composition of the vapour leaving stage n is related to the composition of the liquid leaving the stage *above*. A step-by-step approach to calculate the required number of theoretical stages in the rectifying section can now be introduced.

(1) Knowing the top product composition and hence y_T, calculate x_T from the equilibrium data (for example, figure 6.2).
(2) Insert values for x_T, x_R and R into equation (6.6) to obtain the composition of the vapour leaving the stage below the top one (say y_{T-1})
(3) Re-use the equilibrium data to obtain x_{T-1}.
(4) Re-use the operating line equation, equation (6.6), with the new x value to obtain the composition of the rising vapour from the stage below.
(5) Re-use the equilibrium data and the new y value to obtain the composition of the liquid leaving that equilibrium stage.
(6) Repeat steps (4) and (5) until the liquid composition matches the feed composition.

This simple algorithm shows that in stagewise calculations alternate use is made of equilibrium data and mass balance calculations, which in this case are represented by the operating line equations. Although distillation is a thermal process, the McCabe–Thiele method side-steps enthalpy balances. The key assumption is the one of equal molal latent heats of vaporisation. If this assumption is invalid it is necessary to introduce enthalpy balances in order to calculate the liquid and vapour flow rates.

In order to calculate the number of stages in the stripping section it is necessary to have information about the thermal condition of the feed stream. For a feed at its **bubble point**, the equation of the bottom operating line is

$$y_m = \frac{(F + RD)x_{m+1}}{(R + 1)D} - \frac{(F - D)x_B}{(R + 1)D} \qquad (6.7)$$

The above algorithm can readily be extended to include the bottom operating line. More interestingly, the algorithm can be represented graphically. In figure 6.8 each step represents a stage and the change in composition accomplished by each stage diminishes as the point A (the intersection of the operating line and the equilibrium curve) is approached. If the feed composition is x_{F2}, the slope of the top operating line that has been drawn is reasonable. However if it were x_{F1} (see figure 6.8(b)) this would not be so. With the sizes of the steps diminishing to zero as the 'pinch point' is approached, an infinite number of stages would be required. This slope is related to the minimum amount of reflux. (The slope is, as shown by equation

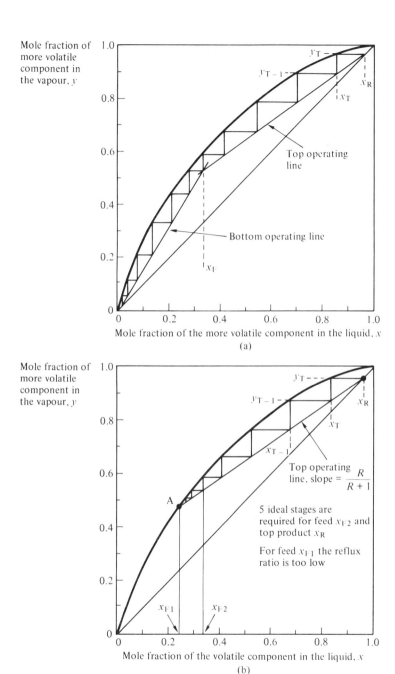

Figure 6.8 *McCabe–Thiele construction: (a) typical construction; (b) illustration of the effect of the reflux ratio*

(6.6), $R/(R + 1)$.) Thus for given values of x_R and x_F there is, for a given mixture, at a fixed pressure, a certain minimum reflux ratio R_{min}, below which the desired amount of separation will not be achieved, no matter how many stages are provided. This is thus a key parameter in the design of any distillation system.

Minimum Number of Stages

The other key parameter is the minimum number of stages. The graphical representation shows that for binary mixtures the number of stages decreases as the slope of the top operating line increases. As R becomes very large the slope $(R/(R + 1))$ approaches unity and the minimum number of stages can readily be found. Figure 6.9 does not include the feed composition because when the number of stages is a

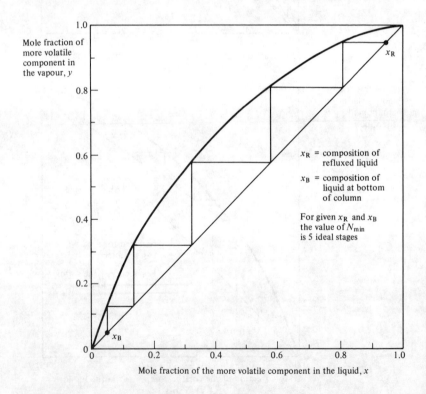

Figure 6.9 *McCabe–Thiele construction for N_{min}. This also illustrates that there is a minimum number of stages for a given separation even when the reflux ratio is infinite*

minimum all of the overhead vapour is returned as reflux and all of the liquid at the bottom of the column is vaporised. Thus there is no feed into the column and no product out. This condition of *total reflux* represents the other limiting condition. The corresponding minimum number of stages will be represented by the term N_{min}.

Balancing Reflux Ratio against Number of Stages

For a given separation a range of reflux ratios can be chosen and the corresponding number of theoretical stages calculated. The resultant data can be plotted and a curve of the form shown in figure 6.10 is generally produced. An increase in the reflux ratio at values near R_{min} gives a marked reduction in the number of required stages. At higher values of the reflux ratio, further increases have a much smaller effect. The economic optimum depends upon a balance between capital and running costs. The latter, which relate mainly to the steam for the reboiler and coolant for the condenser, are roughly proportional to the reflux ratio, whereas the capital costs initially

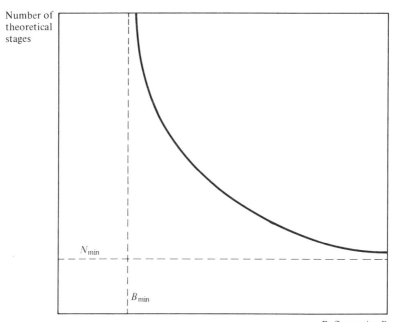

Figure 6.10 *The effect of reflux ratio upon the required number of theoretical stages*

decrease and then increase with increasing R, as shown in figure 6.11. Close to R_{min} many stages are required and the hypothetical column would be tall and thin. With increasing R there is not only a decrease in the number of stages but also an increase in both the amount of liquid descending the column and the amount of vapour rising upwards. Thus although there is a decrease in height with increasing R, the diameter of the column needs to increase, and larger condensers and reboilers are required. Given the shape of the curve in figure 6.10, the sharp reduction in the number of stages is the dominant factor close to R_{min}, but the latter effect dominates beyond a reflux ratio of roughly $2R_{min}$.

Overall, the outcome generally gives an economic optimum of 1.2–1.3 R_{min}. When energy costs were significantly lower than they are now, the economic optimum was reckoned to be somewhat greater at around 1.7 R_{min}. These values are general rules of thumb, and special considerations may dictate values as high as $4R_{min}$.

Choosing and Sizing Trayed Columns

The two most popular types of tray are the sieve tray and the valve tray. The active part of a sieve tray is simply a perforated plate with

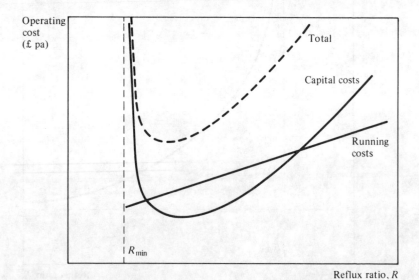

Figure 6.11 *The effect of reflux ratio upon overall operating cost*

hundreds of holes, typically in the range 3–10 mm. Liquid flows across the tray while vapour, which is at a slightly higher pressure on the underside of the tray, is forced up through the holes, and in many systems the result is a bubbly mixture with a froth on top. Sometimes the mixture is predominantly spray. This is particularly true of large-scale columns in which the vapour velocity is high.

A valve tray is shown in figure 6.12. Each hole contains a valve and the extent of opening varies with the vapour flowrate. Thus these trays can be operated over a wider range of flowrates than sieve trays of a similar size. They are said to have a high turn-down ratio which means that they can be operated at a small fraction of design capacity. A figure of 10:1 has been quoted for valve trays, while the turn-down ratio of sieve trays is less than 4:1.

If turn-down is important, a process engineer will have to check his design to ensure that operation at low throughputs is feasible,

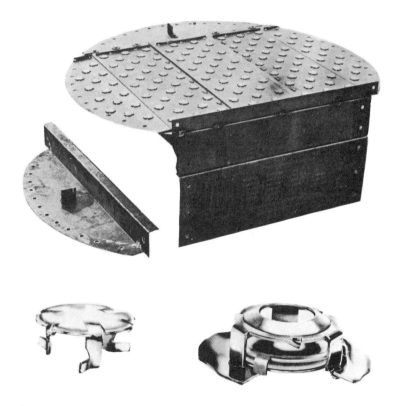

Figure 6.12 *Photograph of a valve tray with sketches of two types of valve*

because the above ratios are just general guidelines. In addition to turn-down ratio, tray efficiency and cost are important with valve trays having the edge in respect of the first factor, while sieve trays are cheaper.

For all types of contacting device, it is necessary to operate within certain limits. For trays, one can readily identify the problems caused by too much or too little liquid, and too much or too little vapour. With excessive vapour flows there will be excessive entrainment of the liquid by the vapour, particularly if the spacing of the trays is insufficient. It must also be realised that only the passage of vapour through the holes prevents the liquid from flowing through them. Thus if the vapour rate is too low, excessive **weeping** will occur. Liquid rates that are too low lead to an opposite problem: liquid is lifted off the plate and the fraction that is entrained (and carried the wrong way) is too high. Lastly, if the reflux is excessive for the size of column, there will be excessive back-up of liquid in the down-comers and liquid will not flow from tray to tray.

The design of a distillation tray is thus seen to be an exercise of keeping the operation within the bounds set by the four problems mentioned above. The tray design is likely to be different for different sections, for example, there will be a significant change in flowrates either side of the feed point. When the assumption of equal molal flows within a section is invalid, there might also be significant changes in flowrate, and hence tray design within a section.

The process engineer (who might be an employee of the operating company or a design engineer with a firm of contracting engineers) will, in addition to designing each tray and specifying the spacing between them, have to complete the column design and specify the materials of construction. The column is in fact a pressure vessel, and design codes are set out in great detail in the appropriate British Standard or its equivalent, which might be an American ASTM code. These will be rigidly adhered to. The challenge will have been in designing the internals which determine the performance and the overall dimensions.

Packed Columns compared with Trayed Columns

The internals of a distillation column or an absorber need not be, and increasingly are not, a series of trays. The packings shown in figure 6.13 can be randomly dumped into a column upon a vapour distributor. The upward-flowing gas or vapour passes through the distributor and follows a meandering path to the top of the packing. This

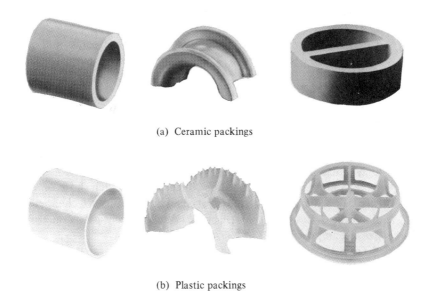

(a) Ceramic packings

(b) Plastic packings

(c) Range of sizes–metallic Cascade Mini-Rings®

Figure 6.13 *Typical random packing materials showing a range of shapes, a range of materials and a range of sizes*

flow is counter-current to that of the liquid which, having been distributed evenly over the top surface of the packing, flows down the packing. There is good contact between the two streams, and matter is transferred between them. The transfer can, for example, be in just one direction, as when a gas is absorbed into a liquid, or two way, which is the case with distillation.

The choice between trays and packing is, in part, influenced by the mode of operation. If a column is operated over a wide range of throughputs, a trayed column is often specified, because the turn-down ratio of most packings is limited to 3:1. However, some select high-efficiency packings can, if the liquid phase is well distributed, achieve higher turn-down ratios which almost match those of a valve tray. With regard to size and cost, packed columns are cheaper if the diameter is less than 1 metre, otherwise generalisations on cost and size are impossible.

Trays are preferred if the removal of heat is important (as, for example, in the absorption of nitrogen oxides in water in the nitric acid process). Cooling coils can conveniently be incorporated into the trays and the high degree of turbulence on the trays aids heat transfer.

With trayed columns, there is for obvious reasons a tendency for the vapour–liquid mixture to froth, while for packed columns the different means of providing contact minimises this tendency. Thus for feeds prone to foaming, packing may well be preferred. The other way of circumventing foaming problems is to dose the liquid with an anti-foaming agent. Packings may also be preferred if corrosion is a problem, because packings are available in ceramic and plastic as well as metal.

Apart from the minority of cases where one of the above factors is decisive, there is a need to consider both trayed and packed columns. Although distillation is introduced in most academic text-books as a classic example of a stagewise separation process, the majority of new distillation columns in the chemical industry (excluding oil refineries) would now appear to be based on packed columns.

ABSORPTION

The advantage of the unit operations approach is that similarities can be emphasised and that quantitative analysis developed in one area can be applied to another. Thus figure 6.3 could have been entitled 'a schematic diagram of a stage within an absorption column', and as a specific example, the absorption of ammonia from ammonia-laden

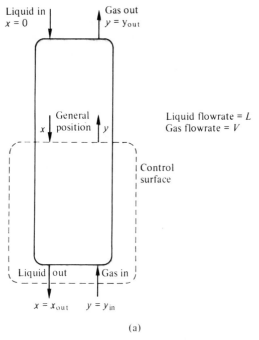

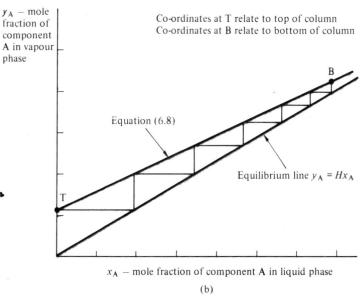

Figure 6.14 *Gas absorption. Note similarity to analysis presented for distillation: (a) illustration of origin of equation (6.8); (b) graphical representation of absorption of a single gaseous component A from an insoluble carrier gas*

air by water could have been chosen. Furthermore, if the concentrations were such that Henry's law applied, the partial pressure of ammonia in the gas would at equilibrium be proportional to the liquid concentration. Writing the expression in general terms with the partial pressure converted to a vapour-phase mole fraction y and the concentration converted to a liquid-phase mole fraction x:

$$y_A = Hx_A \qquad (6.7)$$

Equation (6.7) is represented in figure 6.14(b) together with the operating line which is obtained by taking a material balance around the shown control surface. The equation obtained is

$$y_{in} - y = \frac{L}{V}(x_{out} - x) \qquad (6.8)$$

This lines also passes through the point x_{in}, y_{out}. The amount of absorption achieved in a column of given size is influenced by the ratio L/V. Equation (6.8) is independent of the internals within the column and figure 6.14(b) is the starting point for both simple stagewise and proper packed column design. In the former case a stepping-off procedure of the type introduced for the McCabe–Thiele analysis, starting at point T and ending at point B, would be used. However a packed column is *not* a series of separate stages and a correct analysis is more complicated. It involves a consideration of mass transfer driving forces and film resistances analogous to the concepts encountered for the analysis of heat transfer. Further development is inappropriate at this stage.

EXERCISES

6.1. Engineers often have to make reasonable and *justifiable* assumptions. In distillation analysis it is sometimes necessary to assume 'constant equimolar overflow', a term introduced in the middle of this chapter. By considering various binary mixtures such as benzene–toluene, ethanol–water and ammonia–water, and those of your own choosing, determine for what type of mixtures the assumption of 'constant equimolar overflow' is reasonable.

Although this assumption is still useful when introducing the subject of distillation analysis, it is used rarely in industry today because of the enormous change in desk-top computing power. What do you believe are the main benefits of this change?

6.2. When equation (6.6) is applied to the streams between the top and top-but-one plates, it can be written as:

$$y_{T-1} = \frac{Rx_T}{R+1} + \frac{x_R}{R+1}$$

Show that this is identical to equation (6.1) which was also obtained by performing a material balance. What assumptions have you made?

6.3. Gas absorption calculations require experimental data describing the relationship between gas and liquid compositions under equilibrium conditions. When it is linear, Henry's law applies but there is no universal way in which the data are presented. For the three forms given below, determine the units of H, H' and H''. Are all the values dependent on the total pressure of the system? Can you think of other forms for the equation?

Algebraic form of Henry's law	Units of composition	
	(a) gas phase	(b) liquid phase
$y = Hx$	dimensionless	dimensionless
$p = H'x$	$N\ m^{-2}$	dimensionless
$p = H''c$	$N\ m^{-2}$	$kmol\ m^{-3}$

7 Chemical Reactors

Many types of reactors are needed in the processing industries to cope with the wide variation in production rates and the extensive range of reactions. These include gas-phase reactions in the presence of solid catalysts (for example, ammonia synthesis and oxidation of SO_2 to SO_3) and gas-phase reactions in the absence of catalysts. Examples of these homogeneous reactions are certain chlorinations (such as ethene with chlorine), sulphonation of olefins and the oxidation of nitric oxide.

$$2NO + O_2 \longrightarrow 2NO_2$$

Reactions of commercial significance involving liquids include both homogeneous liquid reactions, a category that includes all aqueous phase reactions, and also gas–liquid reactions. Many oxidation and chlorination reactions involve the bubbling of a gaseous reactant through a liquid reactant. For example, phenol and acetone are produced by (a) bubbling air through liquid cumene (1-methylethyl benzene) to form the stable intermediate cumene hydroperoxide and (b) acidification of the intermediate to decompose it.

$$C_6H_5CH(CH_3)_2 + O_2 \xrightarrow{\text{base}} C_6H_5C(CH_3)_2OOH$$

$$C_6H_5C(CH_3)_2OOH \xrightarrow{\text{acid}} C_6H_5OH + (CH_3)_2CO$$

The absorption of SO_3 and the removal of CO_2 by a chemical absorbent are also examples of gas–liquid reactions. Since the principal aim is to absorb the acid gas, it is generally stated that these are examples of absorption with reaction. However, it should be remembered that the reaction step, which comes after the physical dissolving of the gas at the surface, can control the overall rate of the process. Roughly speaking, the liquid will, in these cases, be saturated with the free gas and more is not absorbed until some is removed by reaction. Thus it is the rate of reaction and not the mass transfer (that is, physical dissolution) which generally governs the overall rate of the process.

With different reaction conditions the free gas may react very quickly with a particular component in the liquid. In this case the actual region of reaction will be confined to the boundary layer and it is the supply of liquid-phase reactants to the boundary layer in the liquid phase, and *not* the kinetics, that will govern the overall rate.

The production of bio-mass is dependent on supplying nutrients (a carbon source, oxygen and all of the necessary trace components) to the liquid surrounding the cells. Thus, oxygen initially present in sterile air is transferred to the liquid from which it is subsequently absorbed and used by the cells. Their rate of growth is often expressed by equations of the form

$$\text{rate} = aC/(b + dC)$$
or $\quad\text{rate} = e$
or $\quad\text{rate} = fC$

where a, b, d, e and f are biological rate coefficients and C is a concentration.

These are not radically different from traditional chemical kinetics and the reader should recognise at least two of the rate equations. Process engineers in the biochemical area need additional background scientific knowledge in subjects such as microbiology, in order to design, in conjunction with others, a safe plant of the right size. This requirement should be compared with other background knowledge of a specialist nature such as petroleum chemistry or detergent science. The need for extra skills is not a dilution of the process engineering element and should not be seen as strange or alien. Apart from sterility the problems associated with contamination, containment and the importance of trace quantities are to be found in many traditional areas of chemical engineering.

A further class of reactions is the gas–solid reactions that include the first stage of the production of zinc from sulphide ores, the reduction of iron oxide in the blast furnace and the combustion of carbon.

$$2ZnS + 3O_2 \longrightarrow 2ZnO + 2SO_2$$
$$Fe_2O_3 + 3CO \longrightarrow 2Fe + 3CO_2$$

Whatever the class of reaction, the process engineer with the task of designing a reactor is likely to find that the chemistry, the kinetics of the given reaction and the required daily production rate are fixed. However he has a number of challenging tasks that include choosing the best type of reactor and its materials of construction, estimating its size and principal dimensions, and determining the optimal operating conditions with due regard to safety, control and any environmental impact.

Batch or Continuous Operation

One of the first questions that needs to be answered is whether the reactor should be operated batchwise or continuously. A batch system gives greater flexibility and generally lower capital costs. Thus it is often favoured for new processes and those with small outputs, particularly if many grades are to be made. Thus batch reactors are common in the pharmaceutical industry and in the manufacture of high-value low-tonnage organics such as dyestuffs. However, for most large-scale processing operations, continuous operation is economically essential. By avoiding the continual filling and emptying of batch vessels, both labour costs and the number of vessels are reduced. Automatic control is also easier with continuous processes, because there is a greater constancy in reaction conditions; one is aiming for steady-state operation instead of being forced, as with batch reactions, to follow and control a continually changing set of circumstances. Lastly, continuous processes generally lead to better product quality.

Two ideal types of reactor will now be considered, after which further consideration will be given to the Case Studies. In both cases it will be assumed that the reaction takes place in one phase only. The system is then said to be **homogeneous**. A reaction is **heterogeneous** if two phases are required in the reactor. The actual location of the place of reaction is immaterial. The analysis of heterogeneous systems will not be considered, but some examples are given in table 7.1.

CONTINUOUS REACTORS

It is worth recalling from early chapters two basic equations which are the fundamental starting point for all unit design, including reactor

Table 7.1 Types of chemical reaction

	Non-catalytic	Catalytic
Homogeneous	Many gas-phase reactions. Some liquid phase reactions, such as neutralisation of acid and alkalis	Many liquid-phase reactions, such as those catalysed by acid or alkalis. Some enzyme reactions
Heterogeneous	Reaction of solids and liquids. Burning of coal. Roasting of ores. Gas–liquid absorption with reaction. Production of bio-mass	Ammonia synthesis. Cracking of crude oil. Oxidation of sulphur dioxide. Gas–liquid absorption with reaction (often promoters added)

design. For any reactant (or product) a material balance can be written. Thus equation (2.6) can, as illustrated in figure 7.1, be applied to reactant A and an element of volume within a reactor. For that element of volume

Amount of reactant A flowing into element of volume	=	Amount of reactant A flowing out of element of volume	+	Rate of accumulation of reactant A in element of volume	+	Rate of loss within element of volume of reactant A due to reaction

(7.1)

When the composition is uniform throughout the reactor (that is, independent of position), the equation can be applied to the whole of the working volume of the reactor. Thus the element of volume equals the volume of reactants. Where the composition is not uniform, the material balance must be made on a differential element of volume and the resulting equation integrated.

For non-isothermal operations, energy balances must be used in conjunction with material balances. The appropriate equation, equation (2.8), can be slightly simplified (because no work is done).

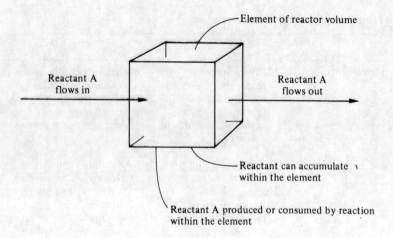

Figure 7.1 *Illustration of material balance applied to component A and an element of volume*

Applied to an element of volume it becomes:

Enthalpy flow into element of volume per second + Rate of production of heat by reaction within element of volume = Enthalpy flow out of element of volume per second + Rate of accumulation of heat within element of volume

(7.2)

Equations (7.1) and (7.2) are linked together because the rate of production of heat depends upon the reaction rate and the rate of loss of reactant. The relationship is mutual and interdependent because the reaction rate is strongly influenced by temperature.

These two equations are the starting point for all reactor designs.

Plug Flow Reactor

The plug flow reactor is an idealised model to which certain types of actual reactors approximate. In many cases a tubular reactor of some sort is visualised but the plug flow assumption is couched in general terms. It is assumed that (a) at any cross-section normal to the fluid flow the velocity is constant, also pressure, temperature and composition are uniform over the cross-section, and (b) there is no mixing between elements in one cross-sectional plane and elements in another adjacent to it. This is illustrated in figure 7.2. The consequence is that the time taken for all molecules to flow through the reactor is the same.

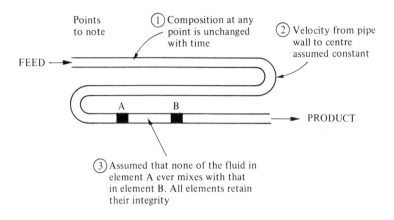

Figure 7.2 *A plug flow reactor: some points to note*

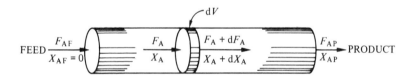

Figure 7.3 *Element of volume* dV: *the starting point for mathematical analysis of plug flow reactors*

With this type of reactor the composition varies from section to section along the flow path, and a differential element dV (see figure 7.3) must be used as the starting point when calculating the reactor volume V. For steady-state the accumulation term is zero. Let

X_A = the fraction of reactant A converted into product
F_A = molar flow rate of A (moles per second)
r_A = rate of reaction of A per unit volume (moles per unit volume per second)

Second subscripts refer to position along the reactor:
F = feed entry point — start of reactor
P = product exit point — end of reactor

141

Then the application of equation (7.1) to the element of volume dV yields for a general position within the reactor:

$$F_A = (F_A + dF_A) + (-r_A)dV \qquad (7.3)$$

that is

$$\text{inflow} = \text{outflow} + \text{amount reacted}$$

The term dF_A can be related to the fractional change in X_A. Since $F_A = F_{AF}(1 - X_A)$, $dF_A = -F_{AF}dX_A$. Substitution into equation (7.3) followed by rearrangement and integration gives:

$$\int_0^V \frac{dV}{F_{AF}} = \int_0^{X_{AP}} \frac{dX_A}{-r_A}$$

therefore

$$V = F_{AF} \int_0^{X_{AP}} \frac{dX_A}{-r_A} \qquad (7.4)$$

The rate of reaction will depend upon concentration and temperature. For isothermal conditions it will be easy to relate the rate of reaction to the extent of conversion and hence perform the integration. For certain simple kinetic expressions this can be done analytically (See exercise **7.1**). Otherwise numerical or graphical integration is required. This is illustrated in figure 7.4.

For non-isothermal conditions the energy balance equation must be solved simultaneously, but in either case the principles behind sizing calculations are simply the application of mass and energy balances. The straightforward example, given in exercise **7.1** illustrates how equation (7.4) can often be simplified.

Continuous Stirred Tank Reactor

The contents of a continuous stirred tank reactor (CSTR) are well mixed and are taken to be of a uniform concentration. Thus there is no need to take balances on differential volumes. The material balance equation for the reactor shown in figure 7.5 is simply

$$\text{inflow} - \text{outflow} = \text{amount reacted}$$

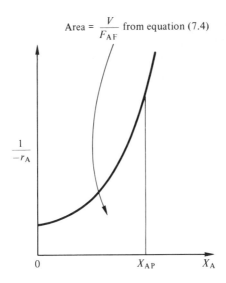

Figure 7.4 *Graphical representation of the performance of a plug flow reactor*

that is
$$F_{AF} - F_{AP} = (-r_A)V \qquad (7.5)$$
therefore
$$F_{AF}X_{AP} = (-r_A)V$$

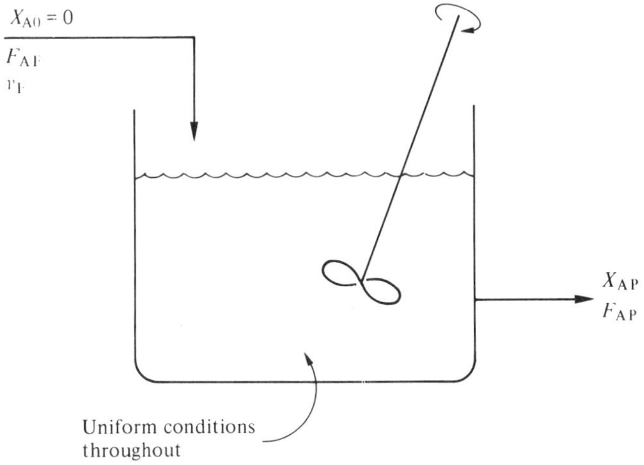

Figure 7.5 *Notation for a CSTR*

If the concentration in the inlet feed stream is C_{AF} moles per unit volume, the total volumetric flow into the CSTR is F_{AF}/C_{AF} unit volumes per second. Let this be v_F. Remembering from chapter 2 that V/v_F is the mean residence time $\bar{t}$, equation (7.5) can be rewritten as

$$\bar{t} = \frac{C_{AF}X_{AP}}{-r_A} \tag{7.6}$$

The mean residence time is thus of crucial importance. Whether equation (7.5) or (7.6) is used there is clearly a simple relationship between size of reactor, reaction rate, extent of reaction and initial concentration. Thus knowing any three allows the fourth to be found directly. In design, the feed conditions and the required extent of reaction will generally be fixed and thus the equation will be used to obtain the required size. (The economic conversion will be dependent upon reactor cost, product sales price and a number of other factors. For fixed feed conditions the equation may be used to investigate how reactor costs, and ultimately profitability, are likely to vary with the extent of reaction.)

In kinetic studies the aim is to calculate the rate of reaction for given conditions. The CSTR is ideal for such calculations because one can check that steady state has been achieved and then without integration obtain the rate of reaction by measuring the throughput and the extent of conversion. As a comparison of equations (7.6) and (7.4) will show, the reaction rate in a CSTR is constant, whereas in a plug flow reactor it varies along the reactor length.

CASE STUDIES REVISITED

The Case Studies are prime examples of heterogeneous reactions and the equations developed above are insufficient because transfer of matter between the phases is as important as the kinetics. Also it is possible in, for example, gas–liquid reactors for one phase to be well mixed and for the other to be in plug flow. Thus in some cases it is no longer possible to assume a uniform flow pattern or state of mixing for the contents taken as a whole. Since it is inappropriate to quantify these effects, the following sections highlight other matters. The exercises at the end of the chapter relate to the equations developed above.

Case Study 1: Sulphuric Acid Production

The first reactor is the sulphur combustion furnace, and a common type is shown in Figure 7.6. Liquid sulphur is dispersed as a fine spray in order to ensure good contact with the air and hence efficient combustion. The furnace consists of a cylindrical steel shell lined with several layers of insulating and refractory bricks. The burner is at one end and the hot products of combustion at about 1000°C are passed directly into a waste heat boiler which produces high-pressure steam from the excess heat. The combustion gases, which contain 10–10.5 per cent by volume sulphur dioxide, are simultaneously cooled to around 425°C, the required temperature for the SO_2 converter.

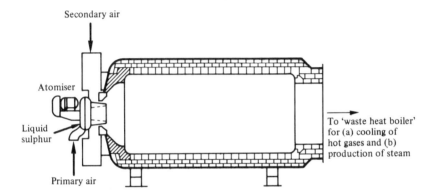

Figure 7.6 *Sulphur furnace*

The process design of a converter requires careful optimisation because there are a number of interacting parameters: sulphur dioxide concentration, feed flow rate, gas temperatures/heat recovery, the number of catalyst beds and the amount of catalyst, gas pressure drop and compression costs. The conventional process of the past is illustrated in figure 7.7. The conversion of sulphur dioxide to trioxide was at best 98 per cent and since the 1970s the principal industrial countries have considered the environmental impact associated with the 2 per cent loss to be unacceptable.

It has been relatively easy to reduce the 2 per cent loss associated with sulphuric acid manufacture by introducing a second absorber before the final conversion stage. Thus after two or three stages of catalytic conversion the reaction gases (see figure 7.8) are passed

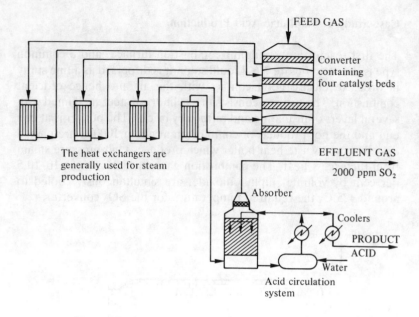

Figure 7.7 *Single absorption sulphuric acid process*

through an intermediate absorption stage which removes most of the product of the equilibrium reaction

$$SO_2 + 1/2\, O_2 \rightleftharpoons SO_3$$

After passage through one or two subsequent catalyst beds the overall conversion efficiency is about 99.7 per cent which means that the atmospheric pollution per tonne of acid produced has been substantially reduced by nearly an order of magnitude (that is, 2 per cent loss down to 0.3 per cent loss). There are now over 200 of these 'double absorption' plants in the world. A typical conversion *versus* temperature curve is shown in figure 7.9.

Process engineers are thus seen to have been able to provide practical solutions to the problems of reducing atmospheric emissions in sulphuric acid plants. They have also achieved a further halving of both wastage and pollution by designing plants that operate at pressure. Increased pressure favours increased conversion because there is a volume contraction on reaction. In France a double absorption plant has been operated at 4 bar for over ten years.

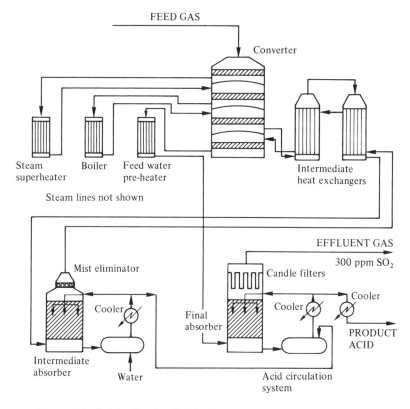

Figure 7.8 *Sulphur-burning double absorption sulphuric acid process*

After much political debate in the 1980s, strict control measures have been introduced for almost all new (and some old) European electricity generating power stations. In any but a local context, this is of far greater environmental significance than the controls on sulphuric acid producing plants because the quantity of sulphurous gases involved is much greater.

By the year 2000, sulphur dioxide and sulphuric acid may be recovered as products from the gaseous emissions of power stations which would bring about a dramatic change in the sulphuric acid market, because the UK power stations, for example, emit, albeit in a weak and complex mixture, more sulphur than the UK sulphuric acid industry uses as raw material. Such a change is possible rather than probable, and proven processes which result in the production of either calcium sulphate or ammonium sulphate are likely to be used to reduce atmospheric pollution.

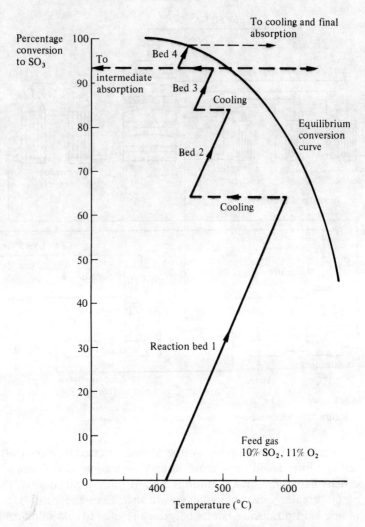

Figure 7.9 *Typical conversion* versus *temperature curve for a 3+1 double absorption converter. There are three catalyst beds before the intermediate absorber and one afterwards*

Case Study 2: Ammonia Production

The conditions in the primary reformer were mentioned in chapter 1 and it will be recalled that the primary steam reforming reaction is overall endothermic. The reactor is a large furnace with around 300 tubes packed with catalyst at a temperature as high as the metal tube will allow. The design of such a system is a challenge for the chemical engineer, who can, with experience, confidently design to within 30°C of the metallurgical limit.

Figure 7.10 is a photograph of the inside of a primary reforming furnace, while figure 7.11 contains drawings illustrating one possible arrangement. The spring supports and the pigtails (coiled small-bore tubing) allow for thermal expansion.

After secondary reforming it is important to reduce the amount of carbon monoxide by way of the shift reaction. A very low level is required because on modern plant any carbon monoxide entering the methanators is converted back to methane to avoid carbon monoxide poisoning of the ammonia synthesis catalyst. This back conversion (a) uses three moles of hydrogen for every mole of carbon monoxide,

Figure 7.10 *Inside of a primary reformer furnace showing the burners and the tubes*

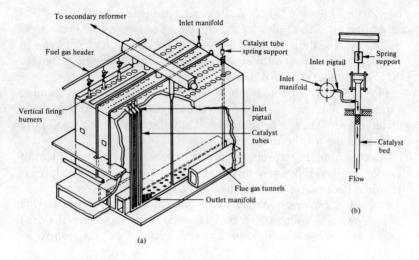

Figure 7.11 *(a) A 'roof-fired box-type' reforming furnace. (b) Detail of top part of a catalyst tube*

and (b) the product methane is an inert in the synthesis loop and must be purged, an action which inevitably results in the loss of more hydrogen. The shift reaction is as follows:

$$CO + H_2O \rightleftharpoons CO_2 + H_2$$

This equation was encountered in chapter 1 as equation (1.6).

The reduction in carbon monoxide level is achieved, as illustrated in figure 7.12, in two shift reactors at relatively low temperatures. The catalysts are selective and the reverse of equation (1.5) is not promoted, thus the methane content remains low at around 0.25 per cent. The catalyst in the second reactor is particularly sensitive to poisoning and sintering. Thus even a few tenths of a part per million of sulphur in the feed reduces catalytic activity irreversibly.

Catalysts have often been referred to, but few details have been given. For heterogeneous reactions of the above type, catalysts are mostly manufactured as the metal oxide on a ceramic support and reduced *in situ* to their active state. (The iron oxide ammonia synthesis catalyst is a major exception, in that it has no support but is simply the oxide with some promoters.) The shape of typical catalysts varies from lumps to pellets to granules, and in addition to being firm

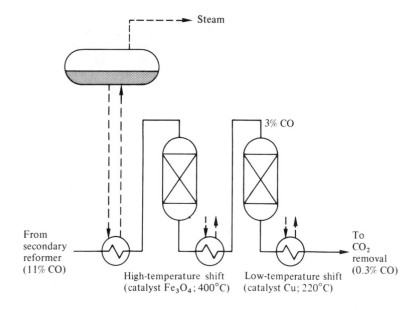

Figure 7.12 *Shift reactors. Used to convert the unwanted carbon monoxide*

they must provide an open structure through which gas can readily flow. Figure 7.13 is a photograph of different types.

The synthesis loop shown in chapter 2 consisted of a recycle compressor, a reactor, ammonia condensation and removal, and a purge stream. A commercial realisation of this scheme is outlined in figure 7.14(a). This may appear to be the only logical order, because it could well be argued that the ammonia should be removed prior to the purge in order to reduce the amount of ammonia lost in the purge stream. However most modern designs, as shown in figure 7.14(b) and (c), locate the purge stream before product recovery. Not all of the ammonia in the purge is lost and an auxiliary chiller and separator, which have not been shown, are provided in the purge circuit to recover as much ammonia as possible.

The difference between the second and third schemes involves the order of the gas-recycle compressor and product recovery. In figure 7.14(b) recovery is after recompression, which gives a higher pressure in the condenser but has the disadvantage that a slightly larger compressor is required because the product is also re-compressed. With the third arrangement, chilling and then refrigerated recovery come before recompression.

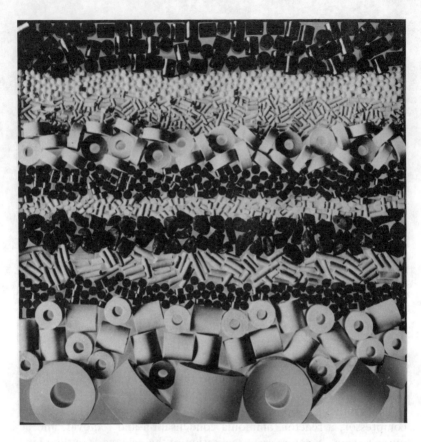

Figure 7.13 *Examples of different types of catalyst*

Many other inter-acting decisions have to be made such as pressure level, inerts level, type of product condensing system and type of reactor. This illustrates the complexity and challenge of process design and hence why it can be seen as a creative activity. There is rarely just one right answer, and it is no coincidence that the words 'engineer' and 'ingenuity' have the same root.

Case Study 3: Related to the Food Industry

During the past few years a lot has been written about 'Biotechnology', and before concentrating upon an example related to food production — the development of a large scale fermentation process for the production of novel proteins — the subject will be placed in context.

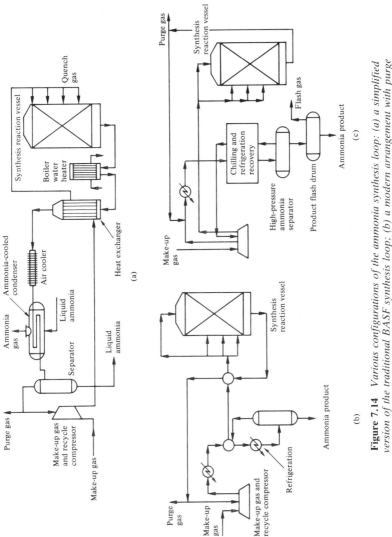

Figure 7.14 Various configurations of the ammonia synthesis loop: (a) a simplified version of the traditional BASF synthesis loop; (b) a modern arrangement with purge and recompression before separation; (c) a modern arrangement with purge and separation before recompression

153

In many ways 'Biotechnology' has been with us for many years, even centuries: the brewing industry, traditional food fermentations, for example, yoghurts and cheese, waste-water treatment and the production of certain antibiotics. These are all important biochemical products. So what is new and exciting? A major change has been the impact of genetic engineering and the wide ranging possibilities that it offers to the fermentation industry. The development of the single cell protein process illustrates this and also indicates the interdisciplinary nature of major research.

In considering the example of single cell protein it is of interest to ask 'Why?' as well as 'How?' By the year 2000, the world population is likely to have increased to 6000 million people. Already many of the people of Africa and Asia are existing on a protein intake below the daily requirement of about 50 g per head. A majority of South American people could fall below this level by the turn of the century. Thus the key question becomes 'Can the protein deficiency be met by conventional means?'

The yield and quality of cereals have been improved and it is also possible greatly to increase the production of soya protein for both animal and human consumption. So, in part, the deficiency can be met by conventional processes, the products of which will definitely form the major part of most people's diets. However, the need is such that new sources may well find a market. Nevertheless, these new sources of protein will be supplements which have to compete with existing supplies.

Although some companies have concentrated specifically on developing novel proteins for human consumption, most research has been channelled into developing proteins for animal feed, and this is likely to be the main path ahead.

Much work has been done since the late 1960s on the development of large-scale processes for the production of novel proteins derived from fermentation processes using yeasts, bacteria and fungi. Although protein produced in this way is often called 'synthetic', this is a misnomer because it is obtained directly from living microorganisms. Many traditional foods such as cheese and yoghurt are produced in a fundamentally similar way, and so the following example, although special, is neither unique nor unrelated to traditional beverage and food manufacture.

The choice of food source is clearly vital. Cost and ease of production are vital and an individual company may well have a preference for one of its own products — either a deliberate product or a waste product. ICI originally tried methane because it was available cheaply from nearby natural gas fields. However not only is it explosive when mixed with air, but since it is only slightly soluble in aqueous solutions there were mass transfer problems. The company turned to methanol, a product in which it is a world leader, because it can be produced relatively cheaply, and unlike methane is suitable for the growth of a single micro-organism.

Coupled with the choice of food source is the choice of micro-organism to grow on that food source. Hundreds were investigated and screened for high growth rate. Suitable organisms were then tested in laboratory-scale fermenters and the most promising strains were screened for pathogenicity and toxicity. The growth medium used methanol as the carbon source, air as oxygen source, ammonia for nitrogen, phosphoric acid and sulphuric acid for sulphur and phosphorus. The organism selected was called *Methylophilus methylotrophus*. One of the few disadvantages of this organism was the low efficiency biochemical pathway used by the organism to take up ammonia. Other microbes were known to possess a better pathway and a successful genetic engineering strategy was devised to identify and isolate a suitable donor gene, to incorporate it into the organism and to isolate unwanted mutants. This was one of the earliest examples of commercially significant genetic engineering.

Fermentation development was also crucial and the laboratory work was followed by the construction of a 37 m^3 continuous culture pilot plant. This was

(a) suitable for scaling up to large-scale tonnage production
(b) capable of high oxygen transfer rates
(c) capable of dissipating the heat of fermentation, and
(d) capable of operating under aseptic conditions.

The nature of the pressure cycle fermenter was mentioned earlier (figure 1.9). Not only does it neatly facilitate mass transfer of oxygen into solution and mass transfer of carbon dioxide out of solution, but less energy is used than that required by mechanical stirrers. Furthermore there is less difficulty in maintaining sterility.

After successful development work, a large-scale plant capable of producing over 50 000 tonnes per annum of product was built. Although a technological success, the plant is not currently operating at full throughput because of intense competition from traditional sources.

Case Study 4: Process Engineering Off-shore

Processing off-shore is kept to a minimum and product upgrading is done on-shore. The one off-shore example of a process involving reaction is acid gas removal. Even this may be done on-shore if the amount of acid gas is moderate. Although the pipeline will have to be a little larger to accommodate the unwanted acid gas, the cost of on-shore purification is so much cheaper than off-shore removal that the savings made will pay for the increased pipeline costs.

While the reactors, which provide the chemical treatment stages, are at the heart of many processes, there are some in which the reactor is an inconsequential unit and others in which it is completely absent. The unit operations, which provide the physical treatment and separation in the preparatory and final stages, are, as seen in earlier chapters, just as important. This particular Case study is a small reminder of these facts.

EXERCISES

7.1. For the special case of constant-density systems, the fractional conversions can be related to concentration. Thus

$$X_A = 1 - C_A/C_{AF} \quad \text{and} \quad dX_A = -dC_A/C_{AF}$$

Also if the kinetic form of the rate expression is simple, then analytical integration is possible. This is so for zero-order, first-order and second-order reactions. This exercise relates to the latter.

If the consumption of reactant A is second order with a velocity constant k, show that equation (7.4) for plug flow reactors becomes

$$\frac{kVC_{AF}^2}{F_{AF}} = \frac{C_{AF}}{C_{AP}} - 1$$

7.2. The alkaline hydrolysis of an ester R.COOR' can be represented by the equation

$$\text{R.COOR}' + \text{NaOH} \longrightarrow \text{R.COONa} + \text{R'OH}$$

The reaction is first order with respect to both reactants, that is, second order overall.

If the velocity constant is 2.0 l min^{-1} mol^{-1} and the feed rate of each reactant into a plug flow reactor is 25 mol^{-1} min^{-1} at a concentration of 0.5 molar, show that the reactor volume required for 95 per cent conversion is 7.6 m^3.

7.3. Show that for a CSTR and a second-order reaction, equation (7.6) becomes

$$\bar{t} = \frac{C_{AF} X_{AP}}{k\, C_{AP}^2}$$

Hence obtain the size of such a reactor if the production rate detailed in exercise **7.2** is to be obtained. Explain in qualitative terms why it is over ten times larger than the plug flow equivalent.

7.4. For the conditions detailed in exercise **7.2**, what conversion is obtained if a CSTR of 0.95 m^3 is used? [*Hint*: obtain a quadratic equation for X_A]

8 Process Control and Safety

The automatic control of processes is widespread throughout the whole of industry and many of today's process operators sit at a computer console checking that many tens of controllers are working satisfactorily. Nevertheless it is worth standing back to ask basic questions concerning the nature of a single control loop such as that shown in figure 1.8 for the cooling of a batch reactor or the one in figure 8.1 for the heating of a stirred vessel with through-flow. Thus the opening sections will be an elementary introduction to process control.

One of the main purposes of control is to ensure safe working conditions, but a plant should not be designed first and controlled second, so consideration is given to safety and its influence on process selection. Techniques for evaluating risk are briefly mentioned, together with some of the approaches that can contribute to an enhancement of safety levels.

FEEDBACK CONTROL

Consider the heating of a living room by means of an electric or gas fire. The occupant switches on the fire, and, for the sake of explanation, selects a middle setting, say 2 kW from a range of 1–3 kilowatts. The occupant may measure the room temperature directly or rely on a perceived level of comfort. A message is then sent either consciously or unconsciously via the nervous system to the brain, where the current room temperature is compared with the desired temperature level (in control language the **'set point'**). If there is a significant difference, action will be taken. The person will adjust the control knob which in turn acts upon the final control element, which in the case of a gas fire will be a valve on the gas line. This scheme should be compared with figure 1.8 and figure 8.1.

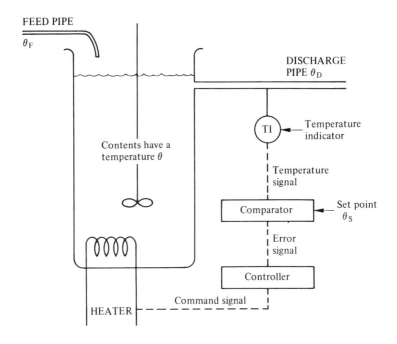

Figure 8.1 *Heater control on a CSTR*

These figures illustrate automatic control systems which are in essence no different from the manual systems that rely upon the action of a human operator. They do the same job. The distinctive characteristic of automatic process control is that the response can be fast and that the controller is pre-programmed to respond in a *precise* manner. The controller can be either pneumatic or electronic. The former have been predominant in the process industries but the latter have found increasing application, especially in new plants. The final control element is often a valve. With a pneumatic controller (which in hazardous environments uses nitrogen rather than air), the output signal can be used directly to operate a pneumatic diaphragm control valve. The output signal from an electronic controller will be either a current or voltage signal and this may be applied directly to a circuit controlling a motorised valve. However these are expensive and a device which converts electrical signals into pressure may be inserted after an electronic controller so that it can act upon a pneumatic valve.

The controller will be fed by a signal from the comparator which calculates the difference between the desired setting and the measured property. If these are temperatures, they could be defined as θ_s (the set point temperature) and θ, the measured temperature at time t. Defining the difference $\theta_s - \theta$ as the *error* the comparator will feed the controller with a negative value if the system is too hot and a positive value if the system temperature is below the set point. Should the output signal of the controller (the signal that is sent to the final control element) include a term proportional to input error signal that is being fed into it? This seems very plausible, and it is interesting to check whether such a controller would perform satisfactorily.

The following analysis is done in respect of the heated well stirred tank of figure 8.1. The energy balance equation (2.8) without the 'work done' term is

Enthalpy flow into unit per second + Heat input into unit per second = Enthalpy flow out of unit per second + Rate of accumulation of energy in unit

Adopting the same symbols as those used earlier for unsteady-state material balances, the above equation is symbolically:

$$m_F C_L \theta_F + q = m_D C_L \theta_D + V C_L \frac{d\theta}{dt}$$

where C_L is the specific heat capacity of the liquid.

As the vessel is well mixed, the temperature of the fluid entering the discharge pipe is equal to the temperature of the contents of the vessel, that is $\theta_D = \theta$. Furthermore if the mass flow rates in and out are steady, $m_F = m_D$. With these two simplifications, the result is a differential equation solely relating temperature of contents θ to time t:

$$V \frac{d\theta}{dt} = m_F(\theta_F - \theta) + q/C_L \qquad (8.1)$$

At steady state, the vessel temperature θ does not vary with time and there will be a heat input q_{ss}, which for a given inlet temperature θ_{F1}, maintains the contents at the set point outlet temperature θ_s. Hence

$$q_{ss} = m_F C_L(\theta_s - \theta_{F1}) \qquad (8.2)$$

The output signal from the controller can now be quantified. When the error is zero it will be equal to equation (8.2). If in addition to this steady-state response there is a response *proportional* to the error, the formula for the controller would be:

$$q = q_{ss} + K(\theta_s - \theta) \qquad (8.3)$$

This is called **proportional control** and because θ is a function of time, q will also be a function of time. It is convenient to eliminate q by combining equations (8.1) and (8.3). This results, after rearrangement, in an equation which states how θ will vary with time:

$$\bar{t}\frac{d\theta}{dt} + \left(\frac{K}{m_F C_L} + 1\right)\theta = \theta_F + q_{ss}/(m_F C_L) + K\theta_s/(m_F C_L) \qquad (8.4)$$

where $\bar{t} = V/m_F$.

The term $\bar{t}$, which has been met in chapters 2 and 7 where it was called the residence time, is once again of importance. In control work it is called the **time constant** because in other control problems a similar term with the units of time, but not related to residence, arises. It is a useful term because the speed at which changes occur can be related to it. After a time equal to $\bar{t}$ has elapsed, the response to a sudden change in inlet conditions is for 'simple' systems 63.2 per cent of the response that will eventually occur. After a time of $2\bar{t}$ the change is 86.5 per cent complete and for $3\bar{t}$ and $4\bar{t}$ the respective percentages are 95 and 98. At this stage a reconsideration of example 2.3 might be useful. The qualification 'simple' is added because the percentages are characteristic of 'first-order' systems which are governed by first-order differential equations. The example was of this type and it would be inappropriate to develop the subject further at this stage.

In order to check whether proportional control is satisfactory, it is necessary to consider how values of K (which is called the proportional gain) affect the response. Purchased controllers have a knob or screw which enables engineers to physically change within limits the value of K and hence the response of a system. While some fine tuning can be done by trial and error, some prior predictions are essential.

If the stirred vessel inlet temperature changes sharply at time $t = 0$ to $\theta_{FI} + \Delta\theta_F$, equation (8.4) can be solved for various values of K.

The solution is:

$$\theta = \theta_s + \theta_F[1 - \exp(-Bt/\bar{t})]/B \qquad (8.5)$$

where $B = 1 + K/(m_F C_L)$.

If $K = 0$, equation (8.5) is very similar to equation (2.5) which came in the section on unsteady-state material balances. Solutions for $K = 0$, $K = m_F C_L$ and $K = 3m_F C_L$ are given in figure 8.2.

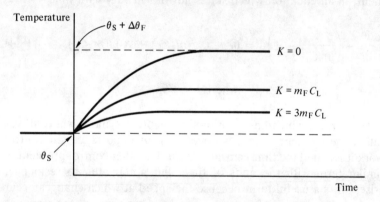

Figure 8.2 *The temperature versus time for various values of* K *after a sudden increase of* $\Delta\theta_F$ *in the inlet feed temperature*

It is interesting to note that for the sudden change in inlet temperature (a type of change referred to by control engineers as a 'load change'), a proportional controller does not bring the system back to the desired set point. The final response is **offset** from the required level. Higher values of K have smaller offsets but, as a little reflection on the implications of equation (8.3) will show, there is a tendency for such controllers to 'over-adjust'. Small fluctuations in inlet temperature such as those shown in figure 8.3 would, even *without* control action, be smoothed by the mass of fluid in the vessel and the response might well be that shown in figure 8.4. A control system with a large value of K would *amplify* the disturbances and in the extreme, the control system might oscillate between having the heater fully on to having it completely off. Not only would the outlet temperature cycle unnecessarily but there would be excessive wear on the control system.

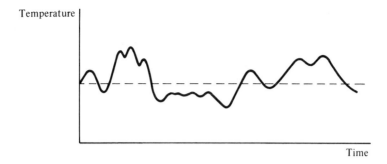

Figure 8.3 *Possible fluctuating behaviour of inlet temperature*

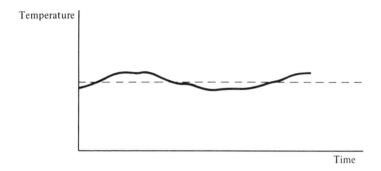

Figure 8.4 *Uncontrolled response to changes depicted in figure 8.3 — note that the thermal capacity damps down the fluctuations*

In order to overcome the offset problem, **integral control** is added to proportional control. The additional term induces a response related to the time integral of the error. For the particular heating problem introduced above, the quantitative expression is:

$$q = q_{ss} + K(\theta_s - \theta) + K_I \int_0^t (\theta_s - 0) \, dt \qquad (8.6)$$

This should be compared with equation (8.3). It can be shown that for any positive value of K_I, the final response is *not* offset from the required level. This eliminates the need to have high values of K and a proportional plus integral controller is clearly superior, at least from this standpoint, to one with only proportional control.

163

However, too large a value of K_I can lead to instabilities and instead of a response such as that in figure 8.5, the oscillations can be of *increasing* amplitude. This is shown in figure 8.6. Clearly, great care must be taken in choosing values of K and K_I.

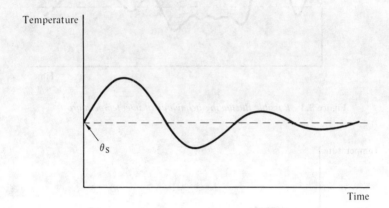

Figure 8.5 *Typical tank temperature* versus *time curve for system with proportional and integral control*

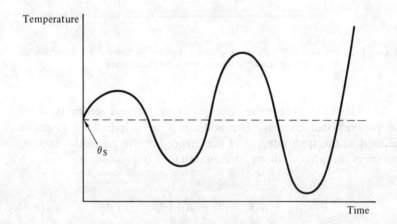

Figure 8.6 *Possible tank temperature* versus *time response for system with large value of* K_I

The above example illustrates some of the ramifications of a typical control problem. Little was said about the measuring element in this particular case, because it would be a thermocouple feeding a thermoelectric voltage back to the comparator within the controller. In some cases, however, the measurement of the desired quantity is difficult.

In the control of a distillation column, the quantity that needs to be controlled is the composition; but this is difficult to measure on a continuous basis and so temperature is invariably used to effect control on a continuous basis, with composition being measured intermittently as a check. Furthermore, the temperature of a liquid does not vary very rapidly with composition and so instead of using the temperature of the liquid at the top of the column as the input to the controller, the temperature of the liquid on a tray in the middle of the top section is controlled at a pre-determined value. Figure 8.7 shows that this is a region in which temperature varies most rapidly with vertical position and so, if the temperature on a particular tray in this region is controlled, then the composition in the whole section of the column is to a large extent under control.

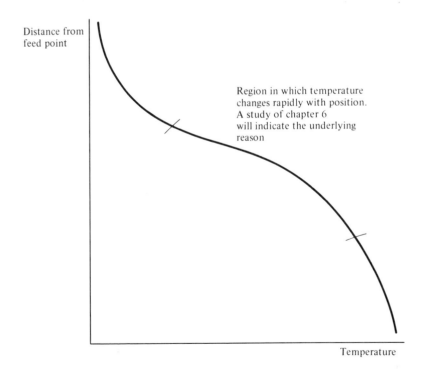

Figure 8.7 *Typical temperature profile down top section of a distillation column*

Currently, in the biochemical industries continuous sensing and measurement is generally limited to pH, carbon dioxide concentration and oxygen concentration. Other quantities would ideally be measured on a continuous basis but the lack of a suitable sensor prevents this. One major exception within a large pharmaceutical company is the use of a mass spectrometer to analyse fermenter vent gas continuously. The results are rapidly obtained and precise. The machine is linked to 64 fermenters and this spreads the substantial capital cost over a number of products.

Biosensors, which use new biochemical techniques to detect specific components in a mixture, are under development and once reliable devices have been produced they should have the potential to transform the biochemical industry. Reliability is vital because otherwise the readings would neither be trusted nor of much use. The devices must work, for example, throughout the whole period of a fermentation of up to six weeks without jeopardising the sterility of the vessel.

In a continuous process the aim is to achieve steady-state, and this facilitates control. With a batch process the conditions are continually changing and the operating strategy can be complex. Computer control can often be justified; the tighter control leads to energy savings, the increased consistency to better products and a reduction in wastage, and, lastly (but not least, despite the fact that it is somewhat intangible) the opportunity to provide alarm trips and interlocks leads to improved safety.

Speciality chemicals such as additives, high value biochemicals and drugs are often produced batchwise. These are an important growth area, and more attention will be paid to such processes and their control systems in the future.

With a computer-based control system one can move beyond the classical single-loop described above, and the action of a control valve can be governed by information fed from a number of points in the process. Since an understanding of both the complexities of chemical processes and advanced control theory is required, this area is at the interface between process and control engineering.

PROCESS SAFETY: AN INTRODUCTION

In order to have a safe process successfully producing to specification the required product, a sound control system is necessary but not sufficient. In addition the plant must be well maintained and the

Figure 8.8 *A modern control room — PVC no. 3 plant, ICI, Barry*

operators must also avoid hazards, not just by the application of common sense, but also by the conscious adoption of safe working procedures.

To illustrate this, consider the following: two relief valves of the same external appearance are removed and sent to the workshop for maintenance. One relief valve is set to operate at a gauge pressure of 1 bar and the other at 3 bar. The set pressures are stamped on the flanges and some may feel that common sense will ensure that they are not interchanged when refitted. However this has not proved to be so, and so the adoption of a more systematic safe working procedure is required. One solution is to dictate that two identification tags clearly numbered, or lettered, be prepared and that on unbolting the relief valve from the flange on the vessel one tag is tied to the valve, the second to the flange.

If unnecessary accidents are to be avoided, additional precautions are required. The original design and indeed the choice of process must be scrutinised to see that it is as safe as is reasonably possible. There is always a certain risk and while it is laudable to quantify the hazards and their consequences, it should be remembered that the perception of risk is often qualitative. Furthermore it is these qualitative aspects which influence society to accept or reject (sometimes irrationally) certain activities. For example, unless society valued the qualitative benefits of personal road travel highly, it would be difficult to explain how it accepts the large number of casualties on the roads.

There has been a tendency in the past both to devalue qualitative aspects, and with regard to communication to stay figuratively behind the perimeter fence. However, more companies are now communicating with society not only indirectly via government agencies, such as the Health and Safety Executive in the United Kingdom, but also to local bodies and even directly to every nearby household. With increasing public concern over waste disposal, toxic emissions and radioactivity, future process engineers will find that their employers value not only their technical expertise but also their ability to communicate both within and without the company at a number of levels.

In the past, many companies also fell into the trap of assuming that if their workers were protected then the public beyond the perimeter fence would automatically be safe. This may be true of fire and explosion hazards but it is obviously not so for toxic releases. Although many of these plants, with thousands of people employed, have often worked for over 10 years without any accident resulting in the loss of a single day's work, the highly improbable but just-conceivably-possible worst cases must be considered. Thus safety studies should identify the hazards, assess the risks associated with each one and specify the necessary safeguards. Such studies consider in detail the possible changes in flows, temperatures and pressures when deviations from normal operating conditions occur. In order to do this, detailed piping and instrumentation (P and I) diagrams which result from the design stage are absolutely essential.

Although such diagrams relate to the operation of plant within a perimeter fence, mal-operation may result in aqueous or atmospheric discharges capable of producing an environmental impact beyond the fence. Thus safety teams who scrutinise P and I diagrams in laborious and systematic ways will not only seek to identify local hazards and their consequences, but also to assess the impact of all conceivable major accidents. The leader of, and maybe a majority on, most safety teams will be by discipline chemical engineers.

An Approach to Safer Design

The constraint of time, the desirability for certainty of production and innate conservatism can mean that although inherently safer processes are identifiable, process plant of a type that proved reliable in the past is built. Some have cautioned that it is therefore essential for those designing a plant to think about 'the plant after next'.

An example is the production of ethylene oxide (oxirane) by the direct oxidation of ethylene (ethene). A few years ago, the existing plants and the plant being designed employed boiling paraffin as a coolant. While (a) boiling-off of a liquid is an excellent way of providing cooling, and (b) the paraffin could operate at the designed pressure, it was clearly a flammable hazard. The company, who had bought the technology, reluctantly did not make a change to the plant under design but resolved to change the process conditions such that the next plant could use high-pressure water. This they did.

At a more mundane level, pneumatic conveying of certain powders can be potentially explosive. If such processes can be modified so that the powders are transported as slurries, this is clearly desirable. Likewise, although there are unlikely to be any sources of ignition within gas–oil separating vessels, it is much safer to start-up by purging the air out with nitrogen and then to introduce the process stream than to start-up with oxygen present. The actual implication of this policy in the North Sea results in the delivery of many compressed gas nitrogen cylinders to the oil platforms in order for them to start-up safely.

In the ethylene oxide example, modifications to the way the process was implemented reduced the hazards. Likewise a modification to the chemical route can be beneficial. For example, instead of reacting A with B and then C to make a compound ABC, it may be relatively safer to make it by reacting B with C and then A to give the same product. The latter avoids the intermediate AB which might be potentially the most lethal chemical compound. An example, relevant to the disaster at Bhopal, is given in figure 8.9. This illustrates *one* objective of all those concerned with design. It can be labelled 'attenuation' — the reduction of hazards in a process which, in terms

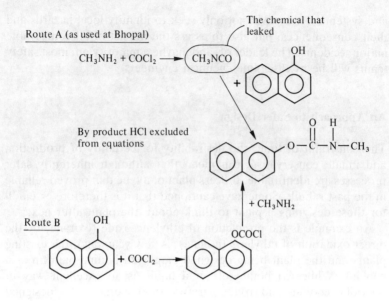

Figure 8.9 *Two routes to the same product*

of chemical route and equipment, stay basically the same. This and two other objectives are illustrated pictorially in figure 8.10.

One of the other objectives is 'intensification' — the reduction of hazards by major reductions in the size of equipment for processes which retain the same chemical route. The classic example is the production of nitroglycerine:

$$\begin{array}{c} H_2CONO_2 \\ | \\ HCONO_2 \\ | \\ H_2CONO_2 \end{array}$$

In the past it was made by the batch process in 1 m³ vessels, but the hazard of handling large quantities led to the development and widespread use of continuous processes. Now the reaction takes place at 48°C in a vessel only 1 litre in size and the products are immediately passed through to a cooler in which the temperature is rapidly reduced to 15°C. The liquid flows by gravity to a centrifuge where the nitroglycerine is separated from the excess acid on a continuous basis. The product is immediately emulsified with a water

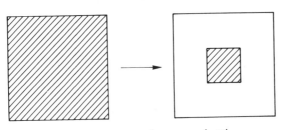

(a) Attenuate — area of concern reduced

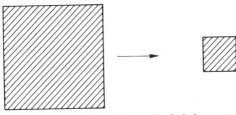

(b) Intensify — reduce scale of whole process

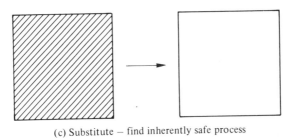

(c) Substitute — find inherently safe process

Figure 8.10 *Pictorial representation of some safety objectives*

jet to form a non-explosive water emulsion for safe temporary storage.

A final route to increased safety is to substitute an inherently safer process for a less safe one. Nylon-6 fibres and plastics are manufactured by polymerising caprolactam, a cyclic amide with a formula

Generally its precursor is cyclohexanone, one of the most widely used chemical intermediates which is also used to make adipic (hexanedioic) acid for Nylon-66 production. Commercially, cyclohexanone can be produced by either phenol hydrogenation or cyclohexane oxidation. The latter can be carried out in the vapour phase, but the preferred route is liquid-phase oxidation with air in the presence of a catalyst at a temperature of 155°C and a pressure of 10 bar. The overall conversion is less than 10 per cent and thus great quantities of cyclohexane have to be separated by distillation and recycled to the inlet where they are mixed with fresh feed. The former route generally involves liquid-phase hydrogenation, catalysed by either palladium or carbon with a conversion exceeding 90 per cent. Not only is the inventory of material much smaller but the process temperatures are below the atmospheric boiling points of the reaction liquids.

The introduction in 1974 of a temporary modification into a well designed and constructed plant, which used the oxidation route, led to a disaster. A bypass line of about 0.5 m in diameter linking two reaction vessels ruptured after 2 months, and over 40 tonnes of cyclohexane, initially at 155°C but with a normal boiling point of 80°C flash evaporated, mixed with air and exploded, killing 28 people and destroying the plant. This disaster at Flixborough, UK, led to a public inquiry. Subsequently, when the plant was rebuilt the alternative process route based on phenol hydrogenation was used.

No plant can be made absolutely safe any more than a car, aeroplane, or home can be made absolutely safe. It is important that this is recognised, for if it is not, processing plant which complies with all of the latest requirements of the day tends to be regarded as absolutely safe and the necessary alertness to risk is thereby reduced. This awareness should, as the previous example shows, be there at all stages, including the conceptual design stage prior to the selection of a particular process route. The challenge for the engineers of tomorrow is, via biotechnological routes or otherwise, to move away from designs based upon the provision of elaborate safety systems that deal with potential inadequacies, and, via attenuation, intensification and substitution, to provide industry and society with processes that provide greater safety.

Conclusion

The fast pace of technological change in contemporary society, which continues to want improved products, processes and increased well-being, ensures that engineering and scientific (both natural and social) skills are in demand. The engineers' plans underpin production of almost everything used in modern life. Their solutions must be acceptable to society not only economically but also environmentally, ecologically and in terms of safety.

The 'process dimension' is being increasingly recognised as a vital component in the analysis of all processes, and there is practically no engineering process in which a chemical or process engineer may not have some involvement. This ranges from research into new processes and products, through development of existing technology and process design, to production which might involve 'trouble shooting' and plant management. In addition to positions of responsibility in these areas, the process engineering perspective is such that with extra training and experience the process engineer can make valuable contributions outside the engineering field in such diverse areas as personnel, economic planning and marketing.

EXERCISES

8.1. Explain qualitatively (a) why high values of the proportional gain K lead to instabilities, and (b) why for any *given* value of K the addition of any integral action potentially makes the system more unstable.

8.2. When discussing the control of distillation columns, the need to control column temperature was mentioned. The temperature is the 'controlled variable'. What do you alter in order to achieve this? (that is, what is the 'manipulated variable'?) A reconsideration of an earlier chapter may be required.

8.3. Describe any accident which has occurred (or *could* occur) in your own home, college or at a place of work with which you are familiar. Discuss ways in which it could have been (or *could* be) avoided.

8.4. Draw up two lists: (a) five basic safety rules which might apply to operators on any chemical plant, and (b) five basic safety rules that those designing chemical plants should remember. Compare the two.

Glossary of Terms

absorption   dissolution of a gas into a liquid
adsorption   bonding of gas or vapour on to a solid surface
anaerobic   in the absence of oxygen/air
azeotrope   a mixture that has, at equilibrium, liquid and vapour phase of the same composition
bacterium   microscopic unicellular organism
boundary layer   region close to a surface over which a major change in velocity, temperature and/or concentration occurs
bubble point   temperature at which a bubble of vapour first appears in a liquid that is starting to boil (depends on composition as well as pressure)
chemical engineering   see page 1 for definition
commissioning   checking of installed equipment, pipelines and control systems prior to production
distillation   application of heat to achieve a separation of the components of a mixture based on differences in volatility between the pure components
entropy   fundamental thermodynamic quantity which increases in an irreversible process and remains constant in a reversible process
enzyme   highly specific biological catalyst; structurally it is a complex protein
eutectic   a eutectic mixture of X and Y will freeze at a fixed temperature to give a solid of the same composition as the liquid; each side of the eutectic composition, either X or Y will start to solidify first
fermentation   a controlled process in which micro-organisms are used to produce desired products
filter aid   a solid that is added to a solid–liquid mixture to improve its filtration characteristics
filtration   removal of solids from gas or liquid

fixed costs costs that are independent of production rate
flaring burning of waste and unrecovered gases
flowsheet a labelled diagram of the unit processes and the main interconnecting pipelines
fouling resistances also called 'fouling factors'; they are thermal resistances generated by the deposition of solid material
fungus are often multicellular and rather complex; they contain no chlorophyl (includes moulds and mildews)
HAZOP Hazard and Operability Studies — a technique in which a combination of systematic rules and engineers' vivid imaginations are applied to proposed designs (or existing processes) to identify hazards
heterogeneous relating to more than one phase
homogeneous relating to a single phase
integral control a method of control which eliminates offset by making the response partially related to the time integral of the error
lactose a sugar found in milk
laminar flow alternative name for 'streamline flow'
Le Chatelier's principle typically paraphased as 'whenever stress is placed on any system in a state of equilibrium, the system will react in a direction which will tend to counteract the applied stress'
liquid–liquid extraction use of a second solvent to remove a selected compound(s) from another solvent
membrane filtration filtration through synthetic membranes
offset difference between final value and the **set-point**
process analysis examination of new processes to assess feasibility or performance assessment of existing processes
process design includes step from initial selection of process, through to the production of flowsheets and the selection, specification and/or design of equipment
process integration study of how new sub-processes can be integrated into an existing process, for example, addition of a cracking unit to an existing oil refinery
procurement obtaining equipment and machinery
proportional control simple method of control in which response induced is proportional to the error
reflux that fraction of condensed liquid that is returned to the top of a distillation column
Reynolds number dimensionless number giving ratio of inertial to viscous forces

sedimentation process of deliberate settling to achieve solid–liquid separation

set point desired level of temperature, pressure, level or some other quantity that is being controlled

shear-thinning apparent viscosity decreases with increasing shear rate

streamline flow a steady flow free of eddies; for flow in a pipe, all fluid particles move parallel to the pipe axis

thermodynamics branch of science concerned with the interchange of heat and work, and the basic principles underlying chemical equilibrium

time constant a parameter with the units of time which may relate directly to a mean residence time or to a simple 'first-order' process (as in chapter 8)

trouble shooting solving urgent production problems

turbulent flow disordered flow containing fluctuating eddies

unit operations techniques founded upon principles of science that are the basic building blocks of a process, for example, distillation, mixing, heat transfer, crystallation

variable costs costs directly related to production rate (equal to zero when there is no production)

visco-elastic describes fluids that exhibit elasticity as well as viscous behaviour (when sudden changes are made to such a fluid it takes a measureable time to reach a new steady state)

visco-plastic describes fluids that only flow when a certain critical yield stress is applied

volatility a measure which compares the amount in the vapour with the amount in the liquid phase; quantitatively the volatility of a substance A is $y_A P/x_A$ where P is the total pressure, and y_A and x_A are the vapour and liquid mole fractions of A

waste heat boiler a heat exchanger in which hot gases are used to boil water (a) to produce steam, and (b) simultaneously to cool the gases

weeping on a distillation tray the liquid should flow *across* the tray and only the upward rising vapour should pass through the holes in the tray; if some liquid does pass through the holes in the tray, then it is said to be weeping

yeast single cell organism found in a wide variety of shapes typically 2–15 μm in size

yield stress that stress that has to be applied before *any* flow occurs

List of Symbols

A	area
B	flowrate of bottoms product
C_f	friction factor
C_p	specific heat capacity
D	distillate flowrate
d	distance *or* diameter
F	flowrate *or* force
H	Henry's law constant
h	enthalpy *or* film heat transfer coefficient
K	proportional gain
K_I	integral action parameter
k	themal conductivity
L	length *or* liquid flowrate
m	mass flowrate
p	pressure
Δp	pressure difference
q	heat flux
R	reflux ratio
R_{min}	minimum reflux ratio
r	radial position *or* radius
T	temperature
t	time
$\bar{t}$	mean residence time
U	overall heat transfer coefficient
u	velocity
u_{av}	average velocity
V	volume flowrate *or* vapour flowrate
Δv	velocity difference
x	mole fraction *or* mass fraction in liquid phase *or* distance
y	mole fraction *or* mass fraction in vapour phase

α	relative volatility
ϵ	emissivity
θ	temperature
$\Delta\theta$	temperature difference
$\Delta\theta_{lm}$	logrithmic mean temperature difference
μ	viscosity
ρ	density
σ	Stefan–Boltzmann constant
τ_o	shear stress

Subscripts

A	component A
D	discharge
F	feed
R	reflux
rad	radiation
s	set point
T	temperature *or* top plate
TOT	total
w	wall

Index

Words printed in **bold** are to be found in the Glossary of Terms on pages 174–176.

absorption 52, 94, 101–2, 132–5
  absorbers 15, 98, 101–2, 145–7
adsorption 92, 94, 101
ammonia, production of 11–16, 148–53
  related examples 35–8
  related exercises 50
  related separation processes 91, 101–3
ASTM (American Society for Testing Materials) 130

Bhopal disaster 169–70
biotechnology 154
boundary layer 68–70, 137
  laminar sub-layer 60
British Standards 130

centrifugation 93, 110
chemical engineering
  applications of 1, 2, 173
  challenge and rewards of 5, 50
  definition 1
  training 10
chemical equilibrium 12, 13, 15, 24
  Van't Hoff laws 11, 12
chromatography 92
commissioning 9
computers 38, 138, 158, 166
crystallisation 93, 110–11

distillation
  choice of pressure 119–21
  general 4, 27, 52, 92, 113–32, 134–5
  minimum number of stages 126–7
  minimum reflux ratio 122–5
  packed columns 130–2
  trayed columns 128–30
  drying 93

electrostatic precipitation 92, 98
emissivity 81–2

energy balance equation 40–2, 142, 160
energy balances 25, 38–44
energy recovery 82–9
engineering contractors 8, 130
enthalpy 39
enthalpy change 85–8
environmental considerations 5, 145, 147, 173
enzyme 62
equilibrium stage 119
evaporation 92, 93

feed-back control 158–65
fermentation 19–21, 103, 152, 154–5
filtration 4, 93, 95–100
  filter aid 97
  filter cake 97
fixed costs 45–7
Flixborough disaster 172
flowsheet 3
fluid flow 52–61
food industry 1, 17, 62, 107, 154–5
fouling factors 74–7
freeze drying 95
friction factor 61

HAZOP *see* safety studies
heat exchanger networks 85–9
heat flow
  linear systems 63–71
  radial systems 71–4
heat transfer
  equipment 16, 71, 79–81
  general 52, 62–90
  resistances 70, 76
  through windows 66–71
heat transfer coefficient
  definition 69
  general 74–7
  overall 70–4
Henry's Law 20, 134–5

179

integral control 163

laminar flow 57
Le Chatelier's principle 11
liquid–liquid extraction 4

material balance
　applied to reactors 139–40
　general 25–38, 50
　in distillation 116, 118
　unsteady state 30–2, 50
　with reaction 32–4
McCabe–Thiele analysis 122–7, 134
membrane filtration 52, 94, 102, 104–9, 112
microfiltration 93, 106–9

Newtonian flow 52–3
nitrogen cycle 11, 12
non-Newtonian flow 52–5
Nusselt number 75

off-set 162
off-shore engineering 22–3, 103–4, 156
oil and gas production 22, 103–4

plate heat exchangers 79–81
Prandtl number 75
pressure—affect on distillation 119–21
pressure drop in pipes 59–61
process analysis 38
process control 9, 17, 158–66, 173
process design 3, 24–51
process dimension 3, 5
process economics 44–9
process flow diagram 8
process selection 109–11, 151–3, 169–72
procurement 9
proportional control 161
protein production 3, 18–21, 152, 154–5
protein recovery 107, 109
pseudo-plasticity 53
purge streams 13, 35–8, 103

radiation 81–2, 90
reactions
　gas–liquid 136–7
　gas-phase 136
　gas–solid 137
　heterogeneous 138, 139, 144
　homogeneous 136, 138, 139

reactors
　ammonia production 14–16, 148–9, 151–3
　batch 138
　continuous 138–44
　continuous stirred tank 142–4
　general 136–57
　plug flow 140–2
　sulphuric acid production 6, 145–8
recycle streams 13, 35–8
reflux 113–16, 118
reflux ratio 121–8
reverse osmosis 106, 109
Reynolds number 57, 75, 96
rheodestruction 55

safety 138, 158, 166–73
safety studies 158
sedimentation 4, 93
separation processes 91–112
shear-thinning 53
shell and tube heat exchangers 79
solvent extraction 92, 95
streamline flow 52, 55–61
sulphuric acid
　heat transfer 62–3, 82
　production 6, 91, 96–100, 145–8
　related examples 29, 33
　related exercises 24, 50
　uses 7
supercritical gas extraction 95

temperature
　driving force 77–9
　logarithmic mean difference 78
　profiles in heat exchangers 78, 84
theoretical stage 119, 126–8
thermal conductivity 63–71
transport phenomena 62
trouble shooting 9
turbulent flow 52, 56–8, 60–1

ultrafiltration 93, 106–7, 109
unit operations 4, 16, 96
U-values 74–7

variable costs 45–7
velocity profiles 57–9
viscosity
　effect on flow 52–5, 95
　effect of protein concentration 54

xylenes, separation of 109–11

yield stress 53

180